D5

On Development

This volume is published as part of a long-standing cooperative program between Harvard University Press and the Commonwealth Fund, a philanthropic foundation, to encourage the publication of significant scholarly books in medicine and health.

ON DEVELOPMENT

The Biology of Form

John Tyler Bonner

A Commonwealth Fund Book
Harvard University Press
Cambridge, Massachusetts
and London, England

Preface

In 1952 I wrote a book called *Morphogenesis, an Essay on Development*. At various times I have mildly contemplated doing a second edition, but the moment never seemed right; the intervening years have been a period of too rapid change. It is hardly certain that the right moment is now, but I have taken the chance. Two main things that affect these pages have happened in the last twenty years. One is of enormous consequence; it is the rise of molecular genetics. In 1953 Watson and Crick saw the "genetical implications of the structure of DNA" and this has opened up a new world. The second is my own increasing awareness of the need to understand development, molecular and otherwise, in an evolutionary context. These are the reasons for writing a totally new book.

In the preface to *Morphogenesis* I confessed to a congenital weakness for biting off more than I can chew. Unfortunately, maturity has not cured me, but rather the condition has become worse. The result is that in this small volume the case cannot be argued in rigorous and extensive detail; it covers so much in so short a space it is bound to be superficial. I am especially sensitive to the fact that much important research in modern molecular biology, though it has a direct bearing on my argument, is not included. I was tempted to make a great list of such experiments and put it in the appendix, but instead I shall merely make a general apology and urge the reader to insert for himself all the many examples (perhaps his own experiments) that would fit. The worry was that if they were all cited, the main point might be lost, and this is a book on ideas, not a comprehensive or critical survey of experiments.

My debt to others for help in preparing this work is very large, and here it gives me great pleasure to acknowledge a few of these kindnesses. Foremost in my thoughts is my good friend J. M. Mitchison, who provided me with ideal working conditions in his Department at the University of Edinburgh during the academic year 1971–72, and was endlessly patient in criticizing ideas and snatches of manuscript and listening sympathetically to generalized author's complaints. It is true to say that almost everyone in the Department of Zoology also helped me in innumerable ways and I thank them each and all.

My colleagues at Princeton have been no less helpful. F. Meins, Jr., allowed me to sit in on his graduate course in Developmental Biochemistry and this was most stimulating to me. He was also very helpful in his critical comments on the manuscript. M. M. Burger, E. C. Cox, A. Newton, A. B. Pardee, and N. Sueoka have, off and on over the last few years, helped me to see my world through the eyes of a molecular biologist. H. S. Horn, E. G. Leigh, Jr., and R. H. Mac Arthur have done the same for evolutionary biology. Furthermore, the depth and clarity of Robert Mac Arthur's way of looking at things (including his critiques of some of my thoughts) were a constant inspiration and guide.

There are many others who helped, of whom I can mention only a few. I am grateful to Walter M. Elsasser for our lively correspondence, which forced me to think more deeply about some rather fundamental matters. Also conversations with Graeme Mitchison were especially helpful to me at a critical moment. One very large debt I owe to E. O. Wilson of Harvard. At the early stages of my work on this book I wrote to him for references on social insects, since I knew he was preparing a book on the subject; by return mail he sent me his completed tome (*The Insect Societies*). It will soon be obvious to the reader how much I have relied on this splendid work. It not only saved me hours of reference tracking, but everything I could possibly want was in one place and described with exemplary clarity and detail. Furthermore, he did me the large favor of reading this manuscript and providing many helpful comments.

I would like to acknowledge, with thanks, the help of Elise Pollock, who did the original drawings.

The writing of this book was made possible through the generosity of the Guggenheim Foundation, for which I am most grateful. I also acknowledge with thanks a special grant from the Commonwealth Foundation Book Fund, which has been invaluable in helping with the expenses involved in the actual preparation of the book.

J. T. B.

Margaree Harbour
Cape Breton, Nova Scotia
June 1973

Contents

On Development

1

Introduction

This is a very general book about developmental biology; it will approach the subject in the broadest way possible, so that the nature of development becomes illuminated by its relation to the other major themes of biology.

But first, on an even more general level, it should be clear that this is not a book about metaphysics. If I say that the cell is a machine or a motor, I merely mean that it can convert one kind of energy into another. If I criticize someone for being a vitalist, it is not (at least in these pages) because I disagree with his religion, but because he has taken a metaphysical position that is inimical to the concept of an open inquiry. However, I am not so naive as to imagine that science can divorce itself from its metaphysical underpinnings, for there are always certain unavoidable assumptions upon which one builds. Of these I would admit only the simplest that would permit one to make observations, to seek regularities in those observations, and to make generalizations from those regularities.

Walter Elsasser has pointed out to me that one of the great confusions in biology has been that words such as "theory" and "explanation" have totally different and very precise meanings in physics; in biology a "theory" is more aptly termed a hypothesis or a model. A biological "explanation" is an especially amorphous concept: it really means any statement that sheds light on a phenomenon, or "makes it plain," as the dictionary would say. One can do this by experiment or observation and reveal a series of prior states or conditions that lead to the phenomenon in question. It is rarely possible to assign a rigorous causal relation to these prior steps; this is more often assumed, and all one can strictly claim is a consistent correlation. But these explanatory descriptions, even for a single phenomenon, can vary enormously. In the case of development we can look at it in two ways: we can look up at it and try to "explain" it in molecular terms, and we can look down at it and "explain" it in evolutionary terms. This scheme has the virtue of simplicity; it is based on a straightforward size scale. Molecules are at the smallest level; development is in the middle range, and evolution or population biology is the largest level. In a sense we shall look at development with both a microscope and the wrong end of a telescope.

This is the same question as the old controversy between the holists, who feel that by looking solely at the parts one can never understand the whole, and the reductionists, who argue that the only way to analyze any problem in science is to understand it in terms of its parts. This latter position has been termed by Hempel and Oppenheim[1] one of seeking a microexplanation, and here we are using the word "explanation" in the loose biological sense; perhaps microdescription would be a more appropriate term. By contrast, the holistic approach would be a macroexplanation or a macrodescription. The important point is that the two approaches to any biological problem are not mutually exclusive: quite to the contrary, they are both essential and reinforce one another.

During the last twenty years developmental biology has become totally transformed, but the change has been for a curious reason. It has relatively little to do with its own progress, which has been steady but slow and sedate. It is almost entirely due to the staggering success of molecular biology and molecular genetics. They have rushed ahead with lightning speed and as a result we see problems of development in an entirely new way. The developmentalist looks anxiously at molecular biology to learn how he might gain deeper insight into their problems, while the molecular biologist feels that now that he has solved the problem of genetics it is time to annihilate the problem of development with the same vigor.

But there has not been an entirely satisfactory meeting of these two minds, and that is in large measure the reason for writing this book. The developmentalist sees the classic problems of embryology in all their elegant complexity and finds himself either unable or unwilling to believe that the biochemist will ever penetrate them with his relatively simple, one-dimensional tools. The biochemist, on the other hand, sidesteps the problem of complexity, often by asserting that one is not asking the right questions. My own feeling is that such a barrier should not exist, and this book is my effort to remove it. I am not a molecular biologist and yet I am totally sympathetic with their views and would like to state the case for development in terms that will have meaning for them as well as for the modern developmental biologist.

There is a third element generally lacking on the part of both the developmentalist and the molecular biologist. This is a willingness to see the problem in an evolutionary context. This is not an idle exercise, but an absolutely essential process. It is only from the evolutionary vantage point that one can see how molecular and developmental biology do indeed fit together.

The case can be stated very simply. The molecular geneticist today believes that all the substances that make up a living organism (with the exception of those that are brought in directly from the outside) are derived ultimately from DNA. The DNA gives rise to specific RNA, which in turn designates specific proteins, many of them being enzymes that control specific organic reactions taking place in the cells. According to this accepted central dogma of Crick, protein cannot make DNA, or other protein for that matter; all the basic instructions for the molecular composition of the organism come from the DNA.

The developmental biologist finds this only a partially useful bit of information. He thinks in terms of what are the instructions that make an egg turn into an oak tree or a frog and he is quite satisfied that they do not all sprout immediately from the DNA in the egg; there are many others, such as the polarity of the egg, the distribution of special chemical components in the cytoplasm, the structure of the egg cortex, none of which seem to have any direct connection with the DNA.

How can this apparent paradox be resolved? What the molecular biologists say of the connection between the substances in cells and DNA is undoubtedly quite correct; the exceptions are likely to be trivial. It is helpful to separate in one's mind three kinds of events that occur in development: the synthesis of substances, the timing of their synthesis, and their localization in space. The DNA role in the synthesis is obvious, but it is less so (and here lies the nub of the difficulty) in the processes of timing and of substance localization. That DNA products are ultimately responsible for localization is probably as generally true as for their synthesis, but, as we shall see, the connection is more remote and the path more tortuous. What we have said so far is that the molecular geneticist is quite correct, yet it is equally certain that the egg or the spore brings with it all sorts of other immediate instructions besides the DNA, instructions which are essential for the development of the new generation.

The obvious answer is that the paradox is an illusion if one looks at it from the evolutionary point of view. Not all the products of the DNA are made *de novo* each life cycle; many are carried from the previous life cycles. Therefore the information for a life cycle does not necessarily all come from the DNA during that cycle, but can come from DNA products manufactured previously. The statement, then, that DNA is responsible for the synthesis of specific substances and their localization can be true only if one adds up numerous life cycles. Both the molecular and the developmental biologist are correct when the matter is put in this way.

It can also be put in another way: there are two kinds of development which are occurring simultaneously: evolutionary development and life-cycle development. The life-cycle developments are in general connected one to another by a single-celled stage, a fertilized egg or some sort of asexual spore, but if one looks at a whole series of life cycles which change through the course of time, then one has an evolutionary development.

This immediately raises the question: why are there two; would not evolutionary development alone be sufficient? The answer lies in natural selection. Selection cannot operate without reproduction and inherited variation, without the making of copies that vary and in turn make copies that replace their progenitors. From the very beginning of life there must have been such a cycling, for, as we shall see, life cycles can never be divorced from natural selection. In the beginning they may simply have involved nucleic acid replication, but as protein came to be produced by and associated with the nucleic acid, eventually an entire prokaryotic cell was constructed and this became the minimum unit or link between one life cycle and the next. This means that since the origin of life the life cycles have become progressively more complex and encompass an increasing number of inner cycles. Multicellular organisms, for instance, will have various molecular cycles, organelle cycles, and even cell cycles. But this is a matter that we shall examine in detail presently; here I want merely to emphasize the life cycle, which leads from one generation to the next.

In any one life cycle, either of a simple bacterium or of a complex vertebrate, the point of minimum size between life cycles is never less than a single cell. It is not possible to make the life-cycle copies, so necessary for and selected for by natural selection, by stripping down to the bare DNA each generation. The amount of DNA is sufficient to accomplish this task, but the mechanical difficulties would be insurmountable. Even if they

5

were not, it would take so much time for each life cycle that the cost of making them would be too great. It is both necessary and efficient to simply have certain processes spread over numerous life cycles, and then one need only strip down to a single cell between cycles.

This means that there is unlikely to be much in the way of change within the cell itself during the entire course of evolution; the change will primarily occur in the life cycle where reproduction and the inherited variation play directly into the hands of natural selection. We know, of course, that this is essentially what has happened. Any genetic change, any mutation, could affect substances that were synthesized at any one time in the great sequence of consecutive life cycles. If it affects the processes that are present in the beginning of a life cycle, such as metabolism or the mechanics of membrane formation, then it is likely to be instantly lethal. If it affects some process that occurs late in the development of one life cycle, such as eye color, then it is likely to be a viable change. There are many intermediate possibilities, each with its own probability of surviving. Therefore evolutionary change is most likely to involve the genes that are acting during the life-cycle development. They are capable and ready for action for the reason that they lie in a cell which, because it is a steady-state accumulation of other genes and gene products, amounts to ideal prepared conditions for gene action. The ultimate result will be increasingly complex life histories or developments, each connected to the next by single cells, and each, through natural selection, finding some way to reproduce effectively in a world of increasing complexity and competition.

I have already said that development will be analyzed in two ways: in one we shall be seeking microexplanations and in the other macroexplanations. Part I of this book is the macro approach: we shall look down on development from the larger problems of evolutionary and population biology. It is not possible to divorce the micro totally from the macro, so there will be a considerable mixture; the parts are separate in terms of their principal emphasis.

Part II seeks microexplanations of development. This is the province of both the experimental biologist and the molecular biologist. The micro elements are of course molecules. How do chemical substances and their reactions manage to go through a series of controlled transformations in time and space that is consistent from one life cycle to the next? How is it that a complex organism, containing a vast multitude of different substances, can develop into a perfectly coordinated, perfectly proportioned individual? The answers to these questions are the hidden treasure we seek. But, as I will try to show, it is the evolutionary view that illuminates the path.

I

The Evolutionary View of Development

2

Cycles and Natural Selection

Because we have such a strong predilection for thinking of organisms as adults rather than as life cycles, the fact that life cycles are an essential ingredient for natural selection generally passes unnoticed.

But first let us sort out the various inner cycles. To begin with, there are two main categories of molecular cycles. (1) Many substances will be synthesized in an organism and later these molecules will be degraded to component parts and then resynthesized. Every aspect of metabolism illustrates this type of chemical cycle. (2) The other kind involves macromolecules such as DNA which replicate by forming templates and then can remain relatively stable for considerable periods of time.

The distinction between these two kinds of chemical cycles is important. Later it will be argued that the elimination of individual organisms is essential for natural selection. This, coupled with the action of nucleases and other degredative enzymes, is the principal means of breaking down macromolecules. But it is equally clear that the molecular stability is necessary in order to carry the inheritance from one generation to the next. Informational macromolecules which multiply by template formation are endowed with both the ability to remain stable and the ability to

11

be eliminated: it is a controlled cycle in which the times of the synthesis and the degradation can be strictly governed. It should be noted that with the action of nucleases the nucleic acids are selectively degraded at different times. Some apparently are eliminated shortly after their formation (for example, *messenger* RNA in bacteria, and possibly much of the RNA found in eukaryotic nuclei), while others, such as the DNA of the chromosomes, remain intact for long periods of time.

On the next level above molecules, various cell organelles undergo cycles. As we shall see later, some of them undergo the immediate disintegration and *de novo* synthesis cycles, while others have growth or replication cycles, like those of cells. (In fact, some organelles, such as basal bodies, can cycle either way, as we shall show.)

The cycling of the cell is one of the most conspicuous of the living cycles. We shall have much to say later about its internal mechanisms; here I merely want to point out the fact that in some instances the daughter cells separate after division, and in others they do not; the latter case is, of course, multicellular growth. Also, cell division is not always by binary fission, but there may be repeated nuclear divisions or mitoses, and sometimes (but not always) this is followed by a cleavage of the cytoplasm to form a series of uninucleate cells. Cell cycling can best be labeled mitosis-to-mitosis cycles. Again these involve a great mixture of template replication and the degradation and resynthesis of molecules and cell organelles.

Finally, there is the life cycle. In sexual organisms this is conveniently thought of as meiosis-to-meiosis (or fertilization-to-fertilization) cycling. Asexual organisms also have a life cycle; in this case it is from bud-to-bud or spore-to-spore (or whatever is the nature of the asexual reproductive body).

Let us briefly examine the degree of coincidence or synchrony of the various levels of cycles we have identified. In unicellular organisms the life cycle and the cell cycle are one and the same. In multicellular forms this is obviously not the case; there can be many cell cycles, and usually the products of these divisions remain attached to one another. But even though there is not a one-to-one correspondence between the two cycles the key processes of meiosis or spore formation involve cell cycles. Or, put another way, in most instances during the life cycle there is a single-cell stage and both the formation of that stage and the resumption of growth when a new life cycle begins involve cell cycles. As Picken[1] pointed out, the life cycle of a multicellular organism is a series of cell cycles in which the cells alternate between adhesive states (growth stages) and nonadhesive states (gamete or spore formation).

The DNA replication cycles are closely synchronized with the cell cycle. However, it is important to note that, although there is one such replication with each cycle, the exact time in the cycle when the replication occurs varies considerably from one cell type to another. Again, some of the organelles, such as basal bodies or mitochondria, may have their cycle synchronized to different degrees with the cell cycle, but the degree varies enormously. As we move down the scale to ordinary molecular cycles the correspondence decreases, although clearly some few particular chemical cycles must be in phase with the cell cycle. The majority are quite out of phase, some because they cycle many times within the span of a cell cycle, and others because substances needed in one cell cycle will have already been synthesized in a previous cycle. (Of course such syntheses could still cycle with the same frequency or period as the cell, but be produced in one cell cycle for a role in a future cell cycle.)

Neither the cell cycle or the life cycle ever goes below a certain size, namely, the minimum cell. In this sense the cell is a steady state which may house many internal cycles but as a whole keeps a well-defined minimum size and structure. This is the mechanism whereby DNA products may be accumulated over a period of time and be available for immediate service in any developmental process. It avoids the necessity of reducing to the bare DNA each cycle, and in that way enormously increases its immediate potential activity: the cell has a large array of enzymes ready for action at a moment's notice, as well as a metabolism machine making energy available in an equally rapid and reliable fashion. Therefore within and among all the cycles there is a steady-state minimum which is the cell.

We are now ready for the main point: why is it that life cycles

13

are crucial for the successful operation of natural selection? The answer lies in the requirements for natural selection itself: as is well known, in order for selection to operate there must be reproduction and inherited variation.

Reproduction is one of the most important processes in biology. In the general sense, it means making copies, making likenesses. This is far more than the mere synthesis of new cell constituents: it means making a whole collection of cell constituents that are, as a collection, a copy of the one that existed previously. The concept therefore involves both time and space. Furthermore, it applies not only to organisms that consist of single cells, but to multicellular organisms, where the individual cells reproduce in growth and the whole multicellular organism reproduces from generation to generation. To make a more precise, general statement about biological reproduction, what reproduces is the life cycle. It is like a motion picture of a continuum that can be recorded by a series of frames from fertilized egg to adult, and then on to the egg of the next generation. It is reproduction in the sense that any one frame at any one point in the cycle will be a copy of the corresponding frame of the preceding cycle and a model for the similar frame of the next cycle. Further, it is also reproduction in the sense that the whole film, the whole cycle, is copied each generation so that temporal and spatial events are duplicated. The larger the organism, the longer the life cycle; nevertheless, both the cycle and each part of it are reproduced. It is, of course, during this life cycle that form changes occur; therefore in a very literal sense our concern with development is a concern with reproduction: development is the copy-making process.

In what was said above I have taken the utmost care to avoid suggesting that the copies are exact, for they do vary slightly. The interesting thing is that just as the exact part of the duplication is inherited, passed on from generation to generation, so are the variations. There may also be developmental accidents which will produce variations that are incapable of being passed on, but this is inconsequential, for such variations will be eliminated as soon as the particular individual dies. It is only the inherited vari-

14

ation that will play a role in natural selection and evolution.

This point is crucial to an understanding of the need for life cycles in natural selection. A somatic mutation is not inherited; only the variations that are in the whole life cycles, that is, the germinal cells, are passed on. Therefore all the lesser cycles we discussed earlier that lie within the life cycle are essential components of growth, of the construction of the life cycle itself, but they may not necessarily be involved in selection unless they happen to be able to pass on an inherited change. However, a mitochondrial or a plastid cycle can, of its own accord, mutate, and if it lies in a germ cell, any of these cytoplasmic organelles may be passed on to the next generation and therefore inherited. But ordinarily the inherited variation will be on the chromosomes of the germ cells (or the asexual spores) and perpetuated in this more conventional fashion. In singled-celled organisms, therefore, the DNA or organelle replication will always pass on the variation from one life cycle to the next, while in multicellular organisms this is true only of the germ cells.

It is impossible to have reproduction, in the sense in which we have defined it here, without life cycles. It is equally impossible to have inherited variation without life cycles, as we have shown. Since reproduction and inherited variation are the prerequisites of natural selection, it follows that life cycles are required for selection.

There is one aspect of life cycles essential to natural selection that is generally overlooked or ignored. It is necessary, in order that selection (either natural or artificial) may operate, to eliminate individuals and not have them existing as adults indefinitely. This is clearly implied in the definition of natural selection, which says that success in selection of any particular genotype is measured by the degree of its contribution to future generations in terms of numbers of offspring. It would be impossible to make any contribution at all if life cycles had no end, no disintegration of the soma. The world would be so jammed with any one species that the resources and space would not permit any offspring whatsoever to appear. Therefore, there has been a selection for life cycles in which the soma exists for a finite time. It is important to examine this point in detail and both review the ideas of previous authors and make a new, additional hypothesis.

It is well known that there is a rough correlation between generation time and the average age at death for any one species.[2] The variation of the life span is large because death may be due to inborn factors that produce senescence, and therefore increase the likelihood of death, or it may be due entirely to environmental conditions which, when harsh, would greatly increase the chance of death. There has been much discussion as to whether or not death-encouraging senescence is selectively advantageous. The current view, progressively refined by Medawar,[3] Williams,[4] Hamilton,[5] and Emlin,[6] is that pleiotropic genes appear that have a beneficial effect in terms of reproductive success at an early period in the life history, and an adverse effect at a later period that results in senescence. The argument is that in terms of natural selection the early favorable effects greatly outweigh the latter deleterious ones, with the result that inevitably the latter will be implanted in the population. Note that in this scheme an increase in reproductive success can be accomplished only by improvements at an early stage in the life history; the later senescent changes are considered inevitable, passive appendages.

Without denying the possibility that senescence may have arisen in this way, I would like to suggest that in addition it may have appeared because, in itself it can be an adaptive trait, especially for those organisms that live in a benign environment and cannot depend upon repeated environmental catastrophes to maintain high mortalities. The argument is that the probability of having a life span of a certain duration is a matter of selective value. This is to say, part of the essential ingredients of a life cycle is not only to reproduce a new cycle, but to get rid of the old one. If this did not occur it would mean that the raw materials to make new copies would soon be exhausted and all the elements needed for effective selection would be buried. A new individual would be competing with his parents, his grandparents, and perhaps many generations further back.

In other words, if an organism can control its mean life span by senescence, it endows itself with the possibility of increased reproductive success. But how can a genetic character that appears either in the postreproductive period or in a period well past the peak of reproductive potential be fixed in a population? One would assume that it would be rapidly lost, since it would be carried only in a small, declining fraction of the population, if at all. To account for its positive selection one would have to assume that it was carried by "kin selection." As with altruism, senescence confers no benefits on the individual, but because its close relatives (such as its offspring) benefit, the genes involved are retained by selection. This is the very same argument that many authors have applied to the postreproductive period of some animals. Because postmenopause individuals help in the care of the offspring, they contribute to the offsprings' reproductive success. These older individuals cannot be selected for directly, but only through genes held and perpetuated by their offspring which they help preserve by their care. The only difference between this situation and senescence is that in the former one is assuring reproductive success by caring for and safeguarding the offspring, while in the latter the success is achieved by getting out of the way so as to avoid using up the resources available to the offspring.

We have discussed a number of conditions needed for selection and none of them are inexplicable in terms of what we know of the physical world. The only difficulty that might remain is how they arose in the first place, that is, how one explains the origin of life, or the origin of metabolizing cells which reproduce and have inherited variations. This is a matter we shall look into closely later; here let me say that this is generally thought not to be an insurmountable problem, and, at least on theoretical grounds, it can be handled with what we presently know of physicochemical principles. The only point I wish to make at this moment is that this does differ from everyday physical phenomena in an important respect. It is usually assumed that once a primitive cell was established, all living forms came from that first prototype which had gained powers of reproduction and inherited variation. Even if it was not a unique event, it must have been an exceedingly rare one. The main evidence for this is that all organisms have essentially the same machinery for metabolism (for instance, the same enzymes) and they all have the same machinery for handling the inherited variation (DNA, RNA). The biggest leap is from prokaryotes to eukaryotes, but even here the majority of the chemical systems are the same for both.[7] No system other than a living organism is endowed with reproduction and inherited variation and therefore only living organisms are subject to natural selection.

We must now turn to the basic question and ask how selection can be so powerful a process as to produce, from our limited random variation, the great diversity of living forms that presently exist. There is nothing haphazard or random about selection. It is a sieve whose properties may change through time, but at any one moment it is fixed. It directly permits certain individuals to be more successful in producing offspring than others; it operates with a statistical efficiency depending upon the sum total of the qualities of each and every one of the variant organisms in a population. The properties of the sieve can be altered by changes in the environment, for indeed the sieve *is* the environment. The environment can change for purely physical reasons such as the climate, or it can change because of the effect of the new populations that are the result of the action of previously constituted selective sieves. But the key point is that at any one moment the sieve is fixed and this rigidity is in itself quite sufficient to produce biological order. The fact that from the pool of variants (and they could even be totally random) one encourages certain types to propagate, and since these types have their peculiarities recorded permanently in their genomes, the character of the subsequent population will change in a specific, consistent way. The inevitable and logical result is order in the evolving populations. It makes no difference how chaotic the variations might be, as long as the sieve consistently lets more of one type through than another, for not only will the character of the population change through time, but the process will automatically result in an increase in order. Furthermore, to do this the sieve requires no magical or demonlike properties: it is the simplest imaginable of mechanisms. Natural selection is a process denied to the world of pure physics; it requires reproduction, and inherited variation, and therefore life cycles, attributes that may have arisen some billions of years ago, and since that time it has been slowly and methodically building new life cycles of organisms, building biological form.

Summary

There are two main categories of chemical cycles: those that involve periodic synthesis and degradation, and those that involve the template replication of stable macromolecules and their controlled degradation, usually by the death of the organism they inhabit. This latter kind of cycle is essential for natural selection, which requires exact reproduction and inherited variation. What is exactly reproduced is the whole life cycle from a single-cell stage (or equivalent), and any mutation will be stable and therefore inherited by the offspring. Life cycles are essential for evolution in every sense, even in the sense that they must include the elimination of adults by death in order to clear the slate for future change. Natural selection is a filter, a sieve that reflects both the physical and the biological environment of any population at any one time; by giving a favorable bias to the reproduction of certain variants, evolution occurs and biological order emerges from essentially random mutation.

3

Reproduction

Here I wish to examine reproduction, one of the two prerequisites for natural selection, in more detail. The conventional meaning of the word "reproduction" is the formation of offspring either by the shedding of gametes that unite in fertilization or by the shedding of asexual bodies, such as spores. Here we are using it in the very much more general sense of making copies. The offspring of an adult animal is, of course, a copy of its parents, but the fact that the embryo of the offspring is also a copy of the embryos of the parents is generally obscured. The conventional use of the word presents even greater difficulties with simple unicellular organisms. A bacterium that goes through repeated cell cycles is hardly the exact equivalent of an adult mammal giving off gametes. Yet the one feature that is fundamental to them both is that they are in the process of making copies, and in both cases they are making copies of the entire life cycle.

Just as the life cycle in its entirety is what is reproduced, it is also what develops. Again the tradition is that the word "development" should be confined to youthful stages, but changes in form can occur throughout the life cycle. The difficulty has been that the most interesting changes take place directly following fertilization; as maturity approaches the number of changes declines, and those that do occur are relatively trivial. The mere fact that the word "embryology" was traditionally used to include all developmental changes emphasizes this point. But again we ask, how does this apply to the life cycle of a bacterium? In that simple case obviously all the stages of the cell cycle are of interest, and not just the stage immediately following division.

Even for large multicellular organisms there is little justification for confining the word "development" to a particular stage. Consider large trees, which continually grow, or many fish and reptiles that do the same thing to a lesser degree, or mammals that grow seasonal antlers; there are so many exceptions that it is reasonable to consider the span of development, like that of reproduction, to extend over the whole life cycle. The difference between reproduction and development is that the former means copy-making and the latter means generating or changing the form. Since the new form is a copy of the old form, the two words describe different aspects of the same processes that occur in the life cycle. The fact that the life cycle repeats itself emphasizes that the instructions can be stored (in a single cell) from one generation to the next. There is a mechanism of converting that stored information into changing form. If natural selection requires reproduction to make copies, it automatically requires development, for it is the life cycle that is copied.

One convenient way to look at life cycles is to think of them as having points of minimum and points of maximum size.[1] Selection has maintained the point of minimum size (either a fertilized egg or an asexual spore) for the optimal handling of variation in the case of sexual organisms and of dispersion in the case of asexual ones. The point of maximum size and all the stages that lead up to it are maintained by selection for their ability to cope in a particular environment. So every stage of the life cycle is adaptive: the minimum-size stage for dispersal or for the handling of variation, and the rest of the cycle for the entire gamut of reasons that give selective advantage to any organism.

If the environment changes, as it indeed has done in different parts of the world over time, it is primarily this period of size increase that will become modified and will adapt to the new

21

environments. The period of minimum size in the cycle changes little: its virtue lies in remaining small and this it preserves zealously. This is the same as saying that, as evolution progresses, not only do we find a wide variety of adult forms to fill the new niches in the expanding environment, but we find new kinds of development as well. There will be a specific selection pressure for a fixed point of minimum size (for effective variation control and dispersion), and a specific selection pressure for a diversity of developments during the remainder of the life cycle.

If we look over this diversity of developments that exist on the earth today, we can see there are certain trends or regularities. Three of these are especially conspicuous and we shall examine each. They are examples of ways in which development has responded to natural selection.

Size increase

During the course of evolution there has been an overall size increase. This does not mean that selection automatically favors large size; in fact, the evidence is good that it favors small size with about equal frequency.[2] Furthermore, in some situations there is selection for no size change at all. But as long as all three possibilities remain open there will inevitably be an upward trend in the sizes of the largest organisms at any one time. Stanley[3] has analyzed this phenomenon in some detail and points out that there are probably severe and sharp physiological and structural limitations when it comes to how small an organism can become, but these problems are less pressing for size increase. Also there is the ecological factor that the small-size niches will be largely occupied, and one way to escape competition and predation is to become larger. That this is precisely what occurred can readily be seen if one compares the maximum sizes, beginning with bacteria of over 3 billion years ago, and following the size maxima right up to the giant sequoia and blue whale of today (Fig. 1).

It is an accepted fact that any major new group arises not from the large species of an ancestral group, but from small species. The examples are numerous: the reptiles that produced the huge dinosaurs arose from small amphibian ancestors, and mammals, which achieved the great whale, sprang from diminutive and inconspicuous reptiles.

The reason for this is probably that the large animals have become overspecialized and are no longer capable of flexibility when it comes to great changes in the environment. Simpson[4] made the point that there are two kinds of specialization: morphological and ecological. It used to be argued that many of the extinctions of animals at the extreme upper end of the size scale were due to morphological specialization: the body became inefficient because the size taxed its design. Nowadays we are far more impressed with the ecological specialization that accumulates with size. Exceptionally big animals become rigidly dependent upon a number of factors in the environment, and any change due either to climate or to other species (either animals or food plants) may easily upset the fragile balance of their niche, with extinction being the inevitable result. Conversely, small species in any particular taxonomic group may possibly (but not necessarily) be less specialized ecologically, and therefore able to adapt to changes with little or no difficulty. It is a curious paradox that natural selection promotes specialization and the subdivision of niches, yet the great leaps forward are usually made by the unspecialized species.

How is it that with each successive major taxonomic group in the history of the earth, the maximum size achieved usually exceeds all previous records? The biggest mammal is larger than the biggest reptile, and so forth (see Fig. 1). Here we assume that there has been a major change in the body-construction characteristics of each new major group and that this change quite incidentally can manage an even larger maximum size than could the ancestral group. A good example may be found in higher plants. There is considerable palaeontological evidence that the first plants with vascular tissue were very small. The invention of this tissue must have been the result of selective forces other than size increase. Yet once this tissue was established, and subsequently when a selective pressure for size increase appeared, it turned out that vascular tissue provided a kind of construction that could support very large trees. So in the course of evolution we find that advances in size increase have been made by an alternating wobble between the selective forces that produce size increase and those that produce major changes in the structure of the organism.[5]

There is a very immediate relation between size and complexity, involving a mechanical dependence of one upon the other; it is impossible to become large and survive in selective competition without becoming complex. A division of labor is essential to support and manage the size. Measuring division of labor or

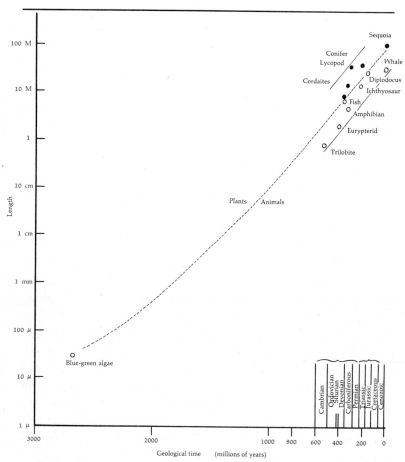

Fig. 1. Maximum sizes (length or height) of animals and plants from different geological eras. (The author is grateful to Drs. Baird, Dorf, Fischer, and Jepsen of the Geology Department, Princeton University, for their help in obtaining the data for this graph.) [From J. T. Bonner, *Size and Cycle* (Princeton University Press, Princeton, N.J., 1965), Fig. 19.]

complexity is next to impossible, but one can make a very rough approximation. Suppose one counted the number of cell types (in itself an approximate and arbitrary task) of the largest organisms of selected phyla and plotted them against the number of cells such organisms contain (Fig. 2). It is clear that there is a correlation between size and complexity, as defined in this way. In order for organisms to become large, they must divide the labor; the two phenonema are inseparable.

In this sense size increase has a direct bearing on development, for large organisms have greater differentiation. More than that, the larger the organism, the longer the period of development, that is, the life cycle. This can be easily shown if one plots the generation times of different organisms against their size. If length is used as a rough gauge of size, there is almost a linear relation between size and generation time (Fig. 3).[6] This increase in duration of the period of development means that there is ample time to do all the complex construction work that is needed to build the large, complex adult. So size increase has a direct effect on the mechanics of development.

This brings us to the question of what the selective advantage of size might be. Why has there been this overall pressure for size increase? First of all, one can discuss selection pressure for size increase only in terms of the small size differences one could expect to appear over a few thousands of years. There is clearly no point in considering the advantage of being as large as a blue whale compared with the size of a whale shark; the two are not competing. Size changes governed by selection must have all occurred in small increments of time. A good example of such size changes is provided by Darwin's finches on the Galapagos Islands. There the evidence strongly favors the notion that one ancestral finch in an environment free of passerine birds rapidly radiated into a number of new species. These species vary in size, so that if all the species are considered their total size range greatly exceeds the presumed size range of the original species, on both the upper and the lower end of the spectrum. By increasing the size range they are making a far more efficient use of the resources of the environment. This is especially clear in those species

25

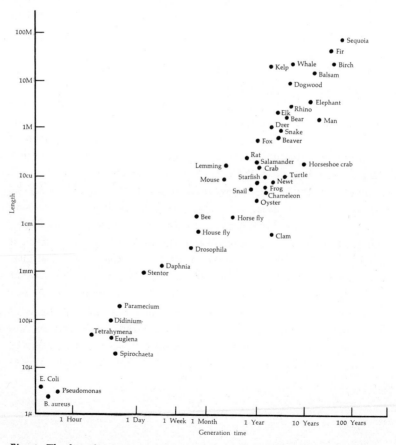

Fig. 2. Very rough estimates of the maximum size (volume and number of cells) of organisms with different numbers of cell types. (For the largest foraminiferan, the number of cells means the number of nuclei, or energids.) [From J. T. Bonner, *Journal of Paleontology*, 42, (Part II of II), 1–15 (1968), Fig. 4.]

Fig. 3. The length of an organism at the time of reproduction in relation to the generation time. [From J. T. Bonner, *Size and Cycle* (Princeton University Press, Princeton, N.J., 1965), Fig. 1.]

that have retained the finch character of seed eating. By the production of larger and smaller seed-eating species, the whole range of seed sizes available in the islands are more effectively consumed. In fact, it can be shown that if two species evolve from one, the size (including the beak size) of one of the new species will be approximately 1.3 times the size of the other, a common phenomenon known as "character displacement."[7] Again it is clear from this example that in these cases there is selection for both increase and decrease in size.

The factors that affect size selection can be generalized into specific functions or activities of organisms. The example we have just given involves feeding mechanisms. Clearly, in competition with other species, the size of an organism might be the object of selection so that the organism can feed more effectively. In the case of the finches it has to do with the food size and there are many other examples where this is so. For instance, in the case of carnivores it is obviously going to be advantageous in some instances if the predator is larger for it might catch its prey more effectively; the larger variants will have a selective advantage. It is equally obvious that this size selection will also operate on the prey, for again the larger forms might be able to put up more of a fight or run faster and therefore be selected. On the other hand, in a particular environment it might be that the smaller prey will be favored because it can hide more successfully. Each case will have its own special characteristics, but clearly size selection is related to feeding.

Size also is undoubtedly in some cases related to success in mating, and since selection is reproductive achievement one can expect size changes. This can be seen strikingly in those cases where there is size competition between males and not between females, resulting in a size dimorphism. In fur seals, mature males may weigh ten times as much as mature females; it is almost hard to believe they are of the same species.

One can make a more general statement of the relation between reproduction and size. There are many instances in which size increase and delayed reproduction are correlated. In salmon, for instance, size appears to be correlated with a number of factors including the length and the swiftness of the current flow of the river that must be negotiated for spawning.[8] Large rivers tend to have older, and therefore larger, spawners. As always, the key is finding a strategy that provides the maximum contribution to future generations in the number of offspring. A large organism can produce more eggs and sperm, but it does so only after a long period of growth (that is, less frequently). Which strategy is best depends upon many factors, both environmental and even internal ones, such as whether or not the organism spawns only once in its life history or repeatedly.[9]

Another kind of function that is size related is the rate of loco-

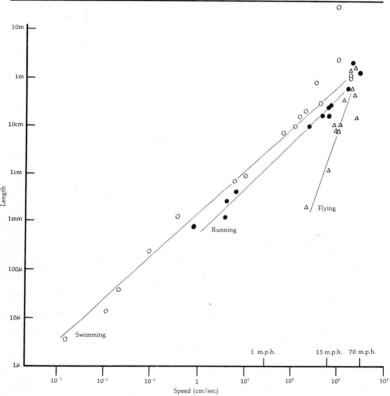

Fig. 4. Maximum swimming, running, and flying speeds of animals of different lengths. The data have been selected for a diversity of types of organisms and for the swiftest examples within each group. [From J. T. Bonner, *Size and Cycle*. Princeton University Press, Princeton, N.J., 1965), Fig. 22.]

motion. This is true for swimming, running, and flying animals (Fig. 4). The larger the animal, the greater is the maximum possible speed of locomotion. Such speed could obviously play a significant role in feeding or mating. It is purely mechanical advantage to be gained from size increase.

Earlier the point was made that one can produce size increase up to the mechanical limit of a particular body plan or construction plan of the organism. Beyond that point the efficiency declines. But as new constructions are perfected by selection for reasons other than size, they might be able to support even greater size without loss of efficiency. This is merely to stress again that each body plan does have its limitations, but new plans are possible. Such large changes are especially conspicuous in the major levels of complexity that we shall presently discuss in detail. The change from the prokaryotic cell to the eukaryotic cell, the change from unicellularity to multicellularity, and the origin of animal societies are all both a jump upward in complexity and a jump upward in size. Furthermore, in the cases of eukaryotic single-celled forms and multicellular forms, within each of these levels there has been a marked increase in size, the latter being conspicuous. Animal societies make a less certain example, although Wilson[10] presents strong evidence that insect societies do reach very large sizes if one measures the whole colony or family.

As one goes up the size scale, especially as one makes some of the major upward jumps that we have described, new properties often emerge. In a sense, we have already said that increases in complexity and in rapidity of movement are two such mechanical, emergent properties; now let me mention a third. It is that as one goes up the size ladder, organisms tend to be progressively more insulated from the environment. A growing bacterium is completely dependent on the presence of food, the right temperature, the right moisture, the right amount of radiant energy, and no doubt other factors as well. On the other hand, a mammal has developed elaborate ways in which it can find food (compared to the limited motility and chemotaxis of bacteria), ways in which it can store energy for long periods of food scarcity, homeothermy to be independent of the external temperature, and a tough integument which keeps down the moisture loss in dry environments and protects the body from a certain amount of harmful radiant energy. The ultimate is to be found in placental mammals, where even the young foetus is kept for an extended period *in utero*, a constant environment quite isolated from the harsh outside world. Size increase has permitted what might be called a progressive internalization, a character that in itself must be of considerable selective advantage.

As with locomotion and complexity, this internalization has a mechanical explanation, at least in its basic mode of operation. Furthermore, this mechanical explanation is the same for all three. It has to do with what D'Arcy Thompson called the "principle of similitude,"[11] and what I have called the "principle of magnitude and division of labor."[12] It is a principle at least as old as Archimedes, and it was put forth with total clarity by Galileo. The volume or the weight of an organism is a cubic function of the linear dimensions. This is a reflection of the total amount of living tissue. However, many of the key activities of an organism vary as the square of the linear dimensions; examples are food assimilation, excretion, heat loss, gas exchange. Diffusion of substances in general takes place through the surface of the organism. Even the strength of a muscle varies with the area of its cross section. For this reason, with increase in size there has been an elaboration of surfaces and levers to keep pace with the increase in bulk. This elaboration is the complexity, the division of labor. The whole relation of size to speed of locomotion can be reduced to these simple principles. Equally obvious is the fact that as organisms become larger they have more inside with respect to outside, making the basic phenomenon of internalization appear automatically. Refinements may follow, such as homeothermy or the formation of a placenta, but even these new adaptations must operate within the limitations of surface-volume relations.

Complexity of the life cycle

There are two kinds of complexity that can appear during development. One is the division of labor we have just discussed;

the other is a complexity in time, a sequence of elaborate events taking place in the life cycle. An extreme example is the appearance of metamorphosis. In insects with complete metamorphosis one finds two developments, one following the other; the development of the larva, and then the development of the adult which begins from the small imaginal discs within the larva.

Before giving further examples, we must mention an important point here that we shall examine later when we discuss the mechanisms of differentiation. The labor can be divided spatially within one organism, or it can be divided temporally so that the entire organism performs first one function and then the next. There are, of course, intermediates between the two, but the complexity we discussed in the previous section was spatial, that is, in one organism the different functions were arranged in space to produce a well-proportioned individual with an alimentary system, a respiratory system, a circulatory system, and so forth.

Consider now a very simple case of temporal differentiation. A spore-forming bacterium will undergo a series of vegetative divisions, and then, depending upon the environmental conditions, it will form a resistant spore. In this case the same cell line will first do one thing and then another, not simultaneously, but seriatim. It should be added that if not all the cells form spores, but only some, then there is also a spatial differentiation or simultaneous division of labor if one considers the whole colony or population of bacterial cells. But as far as any one cell is concerned, it is either a vegetative cell or a spore. This example does have the difficulty that instead of the changes occurring in one cell cycle, the events occur in a sequence of cell cycles, but the complexities are temporal and occur consecutively.

This kind of temporal complexity occurs in a number of other forms besides spore or cyst formation in unicellular organisms and metamorphosis in animals. Another good set of examples is found in plants having an alternation of generations. The fact that one generation is haploid and the other diploid is of far less significance than the fact that they are often morphologically distinct and exist, in some cases, in different environments. In ferns the small haploid prothallus occurs on moist soil, an en-

31

vironment favoring fertilization since the sperm is motile and must reach the egg through a water phase; the large diplophase, on the other hand, can penetrate into the air and is protected from desiccation. It is adapted to catch the sun's energy for photosynthesis, and by firm support of large leaves it can effectively compete with other photosynthetic organisms. Not only are there many other such cases among plants, but they also occur in animals; examples are the alternation of polyp and medusa phases of hydroid coelenterates, and the shifts in the importance and duration of these stages in different species. In insects with complete metamorphosis the larval stage is for eating and growing, while the adult stage is for reproducing.

Another kind of temporal complexity that occurs in the life cycles of some higher plants and animals is directly dependent upon the environment. These have been examined in detail by Schmalhausen[13] for animals; among plants a striking case is that of *Myriophyllum*, an angiosperm whose leaves take on a wholly different form depending upon whether or not the plant is in air or submerged in water. Here it is not merely that there are different ways in which differentiation can occur, but, just as in spore formation in bacteria, the environment may directly influence the switch in pathways.

The question that arises from all this concerns the adaptiveness of these complexities of the life cycle. It is rather obvious that in the examples just given, if an organism can modify its development in direct response to an environmental change, the response will have selective advantage. The more subtle case is in those forms with a metamorphosis or an alternation of generations. Here we are indebted to Levins,[14] who showed that these complex sequences are undoubtedly mechanisms that adapt an organism to a complex environment. Just as one can have both temporal and spatial differentiation in the development of single organisms, so one can have them both in populations of organisms in a patchy, complex environment. Either there may be (to give a specific example) larval and adult forms occurring concurrently because all the insects are not reproducing simultaneously (we gave a similar example with spore formation in bacteria) or the environment can go through temporal changes that correspond to life-cycle changes. In the latter instance the larvae will form in the summer, diapause in the pupa will occur over the winter, and the adults will emerge and mate in the early spring. To complete the picture Levins points out that a phenomenon such as genetic polymorphism is also a response to a complex environment, and this, in our analogy, would be equivalent to spatial differentiation. However, one can also have fluctuations (for instance, seasonal ones) in polymorphic distributions, so that there can be a temporal element to this kind of population complexity also. Finally, it should be pointed out that complex life histories of parasites also serve the same function where the organism copes with a heterogeneous temporal environment by having a series of hosts.

The key point in this discussion is that the temporal complexities of life cycles are adaptive and are successful strategies for coping with complex environments.

Convergence

There is another characteristic of the developmental part of the life cycle that has all the earmarks of natural selection. It is the multiple evidence of convergence in developmental processes.

By convergence we mean that a particular biological function can be fulfilled in a variety of different ways. From the point of view of selection, all that is important is the end result, that is, the performance of that function. The means are of no significance provided certain ends are achieved. The traditional examples come from such functions as flying: the wings of insects, birds, and bats are of vastly different construction, yet they all produce equivalent results. Vision is achieved by the eyes of insects, cephalopods, and mammals, yet these organs were independently developed from primitive photoreceptors. Feeding habits may produce similar structures of independent origin; for instance, the dentitions, of marsupial and of placental saber-toothed tigers appear almost as copies of each other. Examples could be given almost without end. In each instance similar structures have arisen independently through selection in different organisms.

In development there are also an enormous number of processes which can be achieved in different mechanical ways. A good example may be found in the function of growth, which is one of the basic processes of development. If as in most animals, cells do not have hard walls, the synthesis of new protoplasm and cell proliferation can occur almost anywhere, for by cell movement the form of the organism can be preserved. In the case of hard-walled plant cells this is not so and we have growth zones (meristems) or a rigid control over the direction of cell division to produce precise form without any cell motility. Another obvious example would be the physical mechanisms of polarity, that is, the directionality or the symmetry of development. In some cases it is clearly the result of a gradient of some key substance or substances; in others it may be a localized difference in permeability; in still other cases there is evidence that electrical forces play the leading role. We do not yet know the mechanical basis of the majority of instances of polarity in development, but because of the principle of convergence we know that a great variety of mechanisms might exist that all achieve the same result.

The examples given so far involve the broadest, most sweeping aspects of development, but the same principle applies to the smallest detail. In any developmental step there may be a number of ways of achieving it, since selection cares only about the end result. So, if energy is needed for a particular embryonic change, one can expect, and one finds, that different organisms use different fuels. Even such basic processes as fertilization have an enormous number of differences in different species, not only in the morphology of the gametes, but in the chemical events leading to their union. Yet the final result, the ultimate function of the process, is identical in each instance.

From the principle of convergence, we learn that it is to be expected that there will be a great variety of ways in which development mechanically accomplishes certain goals. By looking at development from an evolutionary point of view we learn a lesson that will be especially important to us when in the second half of this book we deal directly with the mechanisms of development. It will help not only to explain the bewildering array of mechanisms, but to emphasize the importance of the common features, the common functions of development which are carried out in so many ways.

Summary

Reproduction (making copies) and development (generating form) are different aspects of the same process, namely, the life cycle. The character of the development is greatly affected by the requirements of the variation system, which are dictated by natural selection. This system must have a single-celled (or small) stage in the life history where the storage is maximal; as the organism enlarges during the rest of the life cycle, the stored information is turned into new form by development. The effect of selection on reproduction is to produce a large number of different kinds of development that are adapted to and compete successfully in particular environments. Three consequences of this selection are of special importance: (1) There will be an overall trend toward an increase in size, which means an increase in the duration of the period of development and of the complexity of the events taking place during that period. It also means a trend toward homeostasis, a progressive insulation from the vagaries of environmental changes and crises. (2) Complex life histories will arise as adaptations to complex environments. (3) Many similar types of development will proceed mechanically in a great diversity of ways. Selection favors certain activities and structures, which may arise independently by convergence, for there is no need that they be mechanically achieved in the same way.

4

Inherited Variation

Besides reproduction, the second prerequisite for the operation of natural selection is inherited variation. The particular relevance of inherited variation to development is basic because what is inherited and what varies is the instructions for development. It is possible to separate the inheritance system into three separate components or functions: (1) storage of the information from generation to generation, (2) variation of the information in a fashion suitably controlled by and for natural selection, and (3) transfer of the information from storage to the production of form in each generation as the life cycle reproduces. In the first part of this chapter we shall be concerned with the first two of these functions; the third will be the subject of the remainder of the book.

Storage in the nuclear DNA

As is well known, the principal means of storing information is in the DNA. Through the rapid and dramatic advances of modern molecular genetics we now know that the DNA contains a code, which is the sequence of bases along the chain of the DNA molecule. There are four different bases and these are read off in units of three; each triplet corresponds to a particular amino acid in the protein that is ultimately constructed. The details of how DNA makes protein with the essential cooperation of RNA need not concern us here, for we are at the moment considering only the question of the storage of information.

It is important to remember that DNA stores information solely for the synthesis of proteins. There has been great interest recently in the question whether or not all the DNA of a cell participates in this function and it would appear that probably a large amount of the so-called repetitive DNA does not. The function or functions of this DNA that is neither transcribed nor translated are uncertain; it has been suggested, for instance, by Walker[1] that it might serve what he calls a "housekeeping" function in the eukaryotic chromosome, such as controlling the folding patterns during mitosis and meiosis. Some of the repetitive sequences clearly are transcribed and translated; for instance, the ribosomal proteins are made by such multiple-copy genes. The storage of most of the developmentally important information, however, presumably takes place in single-copy gene sequences.

It is very difficult to estimate how much storage of developmental information there is in the genome. We know the total amount of DNA per cell for many species, but because different organisms differ so widely in the amount of repetitive DNA we cannot answer this most pressing question. We want to know how many genes exist and what they are used for: we want to know for each of the major groups of organisms what the microbial-molecular biologist presently knows about *Escherichia coli*. Perhaps even more, we would like to know how many genes regulate development, that is, how many genes exist for other functions over and above maintenance and metabolism. Furthermore, of the developmental genes we want to know what proportion are responsible for the production of protein enzymes, and what proportion are responsible for sturctural proteins devoid of any catalytic properties. This is an area of very active research at the moment and we may hope to know some of these answers soon.

Given the raw materials, the necessary elements, all the compounds inside a cell are synthesized by enzymes. Even in the case of proteins themselves, enzymes are necessary for pushing each step of the DNA-RNA-protein chain of synthesis. This means that protein enzymes are responsible for the synthesis of all the carbohydrates, fats, proteins, and the plethora of other substances that exist within a living and developing organism. Therefore the stored information in the DNA can have far-reaching effects beyond the construction of proteins, but this is a subject to which we shall return when we discuss the action of genes in development.

Storage in other substances

It is self-evident that in living organisms storage of information in the DNA alone is insufficient to produce a life cycle. Were this all that existed after fertilization, there would be no possibility of protein synthesis or the synthesis of all the other necessary substances, even if the elements for organic compounds were provided. Since life cycles are circular or repetitive, it depends upon where one begins. If it is at the formation of a zygote or a spore, DNA polymerase is needed for development, as well as RNA polymerase and the other enzymes at the level of the ribosome that participate in protein synthesis. Obviously the first synthesis of proteins in the fertilized egg or germinating asexual spore requires preexisting enzymes.

The very same is true for viruses, even though they are made up largely of nucleic acid. But they have protein as well, which they use for such processes as attachment to and penetration of the host: in some cases the protein is structural; in others it may be an enzyme. Viruses are pertinent for two reasons. (1) Some consist of RNA rather than DNA, and now it is generally recognized that in some mammalian systems RNA can carry the code and synthesize new DNA with an altered code: the so-called reverse transcriptases. This means that RNA can assume a primary storage role similar to that of DNA. (2) More important is the fact that a virus cannot use the code of its stored nucleic acid unless it parasitizes a living cell. It uses the protein-making machinery of a living cell to manufacture viral proteins. It slips its code into the preexisting protein factory.

A virus points up a very fundamental aspect of the storage of information for development in all living forms. Not only are the DNA and a set of appropriate proteins needed, but also the energy machine, the metabolizing cell. The reason for this is largely that only through metabolism is it possible to capture energy and drive the chemical reactions necessary to produce the nucleic acids, the proteins, and the other cell constituents. These reactions cannot occur without exogenous energy, and the cell motor is ideally suited to provide this energy necessary for synthesis. All the constituents that are essential for the proliferation of a virus are ones that derive from a cell in active metabolism. The virus parasite usually kills the host cell after it has accomplished its life cycle and goes into its storage phase again. But it stores only part of the information necessary to begin a new cycle; it plans to plunder the rest from a fresh metabolizing cell.

Therefore part of what is stored is the running motor itself. This automatically includes a vast number of other materials which are the key gears of the metabolism machine. But the concept goes much further than the need to store a whole stockpile of substances necessary for metabolism. The new life cycle is dependent not only on the presence of these substances, but on the fact that they are actively metabolizing. Like so many totally obvious things, this can be easily overlooked. Besides DNA and a large number of other substances, besides the motor of metabolism and all its parts, one also inherits the running of the motor. It is an eternal flame, for a flame is also an energy-converting process. The only difference is that a metabolizing machine can pause (as with dormant seeds or spores) but, because the parts of the motor are all in place, favorable conditions permit it to begin again, like an automobile with a self-starter.

Storage of structure
Only one further aspect must be added to complete our picture of what is stored. We are really always referring to the point of minimum size in the life cycle: the zygote or the asexual spore. This is the point where the maximum amount of storage is required and the point that is invariably followed by a period of development and size increase. The minimum size is one cell.

This is the smallest unit that can contain DNA, the associated proteins, and the running or self-starting motor. Therefore in a very strict sense the whole cell is the storehouse, even though its constituents, particularly the DNA, store different levels of information in different ways.

The cell has its multitude of substances arranged in a well-organized series of structures, and these structures are in themselves part of what is stored. If we confine ourselves to eukaryotes, the cell membrane is always present and can apparently be extended as the cells divide during the size increase of development. We are assuming here that membrane must form from previous membrane and that what is inherited is the basic pattern of the structure, like a seed crystal. Of course it is quite possible, and this no doubt applies to many of the structures of the cell, that cell membrane can also arise *de novo,* and that the fact that we always see it arise from preexisting membrane may be irrelevant; in order to get membrane, storage may not be required.

Let us turn to some of the more organized organelles, which probably do not arise *de novo.* Mitochondria and plastids, for instance, are stored and inherited as intricate structures (Figs. 5 and 6). Furthermore, it has been known for a long time that since these structures are free in the cytoplasm they can be inherited independently of the nucleus. In maternal or cytoplasmic inheritance in plants the pollen does not contain any plastids and therefore all the plastids of the offspring come from the egg. If the pollen comes from a plant with a different plastid phenotype (for example, a yellow instead of a green color), this will be lost in the offspring, for they receive only the maternal plastids.

Mitochondria are found in gametes; if a mitochondrion is altered in some way, this alteration can be passed on from generation to generation. The petite mutants of yeast have deficiencies in their cytochrome system, and if a cell has nothing but the mutant mitochondria, this will have a drastic effect on the appearance of the colony, which grows poorly. Here there is the interesting further point that sometimes nuclear genes affect the expression or the multiplication of these variant mitochondria,

40

so that cytoplasmic structures can also be controlled by the nuclear DNA.

Both plastid and mitochondrial inheritance have become matters of special interest with the discovery that each individual organelle of both types contains DNA and RNA. Furthermore, it is clear that they have the necessary machinery, such as their own

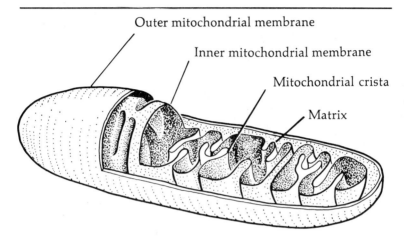

Fig. 5. Structure of a mitochondrion. [From W. R. Bowen, *Experimental Cell Biology* (Macmillan, New York, 1969), Fig. 3–13D.]

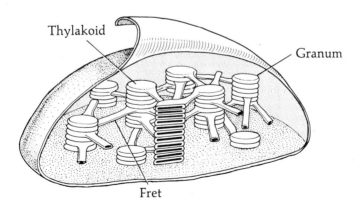

Fig. 6. Structure of a plastid. [From W. R. Bowen, *Experimental Cell Biology* (Macmillan, New York, 1969), Fig. 3–16C.]

ribosomes, to make proteins. There is evidence that these organelles make a few of their structural proteins; the rest are made
by the nuclear machinery. The fact that the structures of both
the DNA strands and the ribosomes of chloroplasts and mitochondria resemble those of bacteria have suggested that they may
originally have been symbiont prokaryotes.[2] This is also consistent with the fact that plastids and mitochondria do not arise
de novo: they self-duplicate. For this reason they may be considered major storage sites of structure, containing their own
protein-synthesis mechanism that perpetuates itself during
growth.

Another cell structure important to this discussion of storage
is the basal body, which is thought to be identical to the centriole
(Fig. 7). Unfortunately, there remain a number of key unknowns

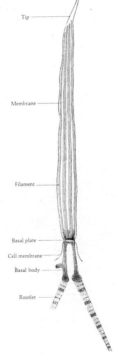

Tip

Membrane

Filament

Basal plate
Cell membrane
Basal body

Rootlet

Fig. 7. Longitudinal section of a cilium, showing its attachment to a basal
body. [From P. Satir, *Scientific American* (February 1961), p. 110.]

42

about these curious structures; for instance, in most cases they appear to be clearly self-replicating, but in the transformation in *Naegleria* from the amoeba to the flagellate state, there seems to be a *de novo* appearance of the centriole.[3] An even clearer case of the resynthesis of basal bodies has been shown by Grimes[4] in *Oxytricha*. All the basal bodies disappear as the organism encysts and they are remade completely upon germination from the cyst.

It is interesting that the centriole is concerned with mitosis and the formation of the spindle, while the basal body is part of the cell cortex and is the root of a cilium or a flagellum. It has been known for a long time that the two are connected, for in some forms the centriole will divide in two, and one will wander to the surface and sprout a flagellum. Structurally they are identical; the only reason for the difference in their names is their position in the cell. Among other things this relation shows a structural connection between the inside and the surface of a cell. Even though it is not always clear, in the majority of cases basal bodies are known to self-duplicate and therefore provide another example of cytoplasm-structure storage.

In ciliate protozoa, where the basal bodies are lined up in well-oriented rows, information is stored not only in the structure of the individual units, but also in the whole pattern of the units in the cell cortex (Fig. 8). These organisms are peculiar in that they do not revert to a simple set of gametes that then develop into complex adult structures. In most cases they retain much of their adult cortical structures at all times and achieve fertilization and recombination by two individuals of opposite mating types pairing or conjugating; while attached they exchange haploid nuclei and then separate. Each partner has a new zygote nucleus with which it rebuilds a new macronucleus. The fine structure of the cortex remains constant and the two new offspring still wear the shells of their parents. In this example the entire cortex is a storehouse of structural information which is passed on directly either through conjugation or through cell division. This does not mean that gene changes cannot ultimately control the appearance of new cortical patterns, but how this occurs is less well understood than the direct cortical storage and transmission. It should be added parenthetically that there are other ciliates, for example the suctorians, in which the neat rows of basal bodies are realigned in each generation; the pattern is not directly inherited, but only the potential to achieve the pattern, which is then fulfilled during the course of development. We shall return to this example when we discuss the whole matter of self-assembly.

The question of whether or not polar granules are part of the basic information passed on from one generation to the next is particularly interesting.[5] There are a number of animals, especially some insects, nematode worms, and vertebrates, in which

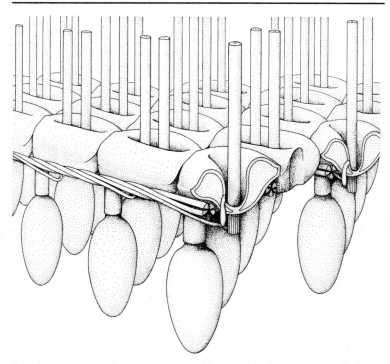

Fig. 8. Diagrammatic reconstruction of the cortex of *Paramecium*. Each pair of cilia emerges from the center of a polygon; the pairs of inflated alveoli defining the polygons are shown in section at the right-hand edge of the stereogram. Parasomal sacs are shown adjacent to the cilia in these polygons. Resting trichocysts alternate with the polygons in longitudinal rows. Kineto-desmal fibers form loose cables paralleling each kinety. [From J. O. Corliss (redrawn from Ehret and Powers 1959), *The Ciliated Protozoa* (Pergamon Press, Oxford, 1961), VI–2C.]

very early in the development certain cells are laid aside to become the germ cells that give rise to the next generation. This is
the direct manifestation of Weismann's old ideas of the separation of the germ plasm and the soma. The germ region is associated with special granules, seen particularly clearly in insects
(Fig. 9). If this region of the egg is either removed by surgery or
destroyed by a small beam of ultraviolet light, the egg will develop into a sterile but morphologically normal adult.

The granules themselves are known to be a mixture of protein
and RNA; it has been suggested that they might be messenger

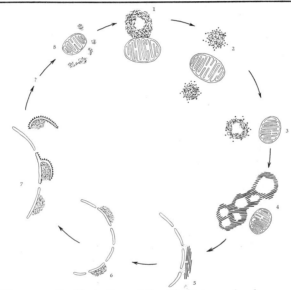

Fig. 9. Diagrammatic illustration of the life history of polar granules. Beginning in the mature oocyte (1), the polar granules are typically attached
to mitochondria, then after fertilization (2) they become free, fragment, and
have ribosomes attached to their periphery. After pole cells have formed, the
number of ribosomes attached to the granules decreases, and reaggregation
of the fragments occurs (3) until the completion of the blastoderm stage,
when large aggregates have formed and ribosomes are not associated with the
granules (4). Subsequently, during the formation of the embryonic gonad,
the large aggregates again fragment, become localized along the outer nuclear envelope (5 and 6), and remain in this location until the next cycle of
oogenesis (7). At this time they are frequently associated with rough endoplasmic reticulum. Although the formation of polar granules during oogenesis is not clear, it is possible that this same fibrous material becomes localized
in the posterior region of the egg (8) and becomes organized into the typical
polar granules of the mature egg (1). [From A. P. Mahowald, in J. Reinert
and H. Ursprung, eds., *Origin and Continuity of Cell Organelles* (Springer-
Verlag, New York, 1971), Fig. 6.]

RNA in storage. They clearly exist in the egg and can be followed through a series of transformations accompanying the formation of germ cells. The question of whether they remain intact and move to the polar region of the eggs of the next generation is still unsettled. They could be re-formed during oögenesis. But in any event they are certainly part of the stored information ready for the next generation. There is no answer at present to the very interesting question of what kind of information they possess: do they stimulate the formation of germ cells or do they protect and prevent certain cells from turning into anything else? Of all the cells in the body the germ cells obviously have the most far-reaching potency.

There are, of course, other structures that we could discuss, but the main point has been made. Besides the central nuclear DNA storage, there is a multitude of others within the cell and the sum total of all of them is required if we are to have successive life cycles of any organism. Finally, the ultimate key to successful storage is the storing of the running motor of active metabolism.

There is, however, one point that must be clearly understood. All the structures we have mentioned, and all the substances that are involved in the energy machine (with the exception of those that are directly pulled in from the environment), are ultimately determined by the DNA (both nuclear and cytoplasmic) through the RNA-protein route. All we are saying here is that at the point of minimum size in the life cycle much of this DNA information has been brought out of the ultimate DNA storage and has already been realized in the form of membranes, organelles, and all the cytoplasmic components that make up a cell. Even though each cytoplasmic structure may be, directly or indirectly, in the DNA code, there is a strict limit, at the single-cell stage of the life cycle, to how much information can exist solely in the DNA and how much must already have been read out or synthesized from the DNA in previous cell cycles.

Storage in animal societies

Animal societies may be roughly divided into two extremes: those such as insect societies, which seem to inherit the majority

if not all of their behavior patterns (in other words, all the information is stored in the egg), and those such as mammal societies,
which bypass the egg with many significant bits of behavioral
information and pass them directly from parent to offspring by
teaching and learning. Most social organisms have a mixture of
the two methods of information transmission, with social insects
at one end of the scale and *Homo sapiens* at the other.

The evidence for the lack of information transfer by learning
in insect societies comes from work in which the callow adults,
emerging after pupation, are isolated from other, older individuals. The result is a slight delay of 1 or 2 days in assuming their
duties as workers or soldiers, but ultimately the tasks are performed in a professional fashion.[6] In other studies the pupae of
one species are put in the nest of a closely related species, yet the
hatching workers retain their original, species-specific behavior
patterns in nest building and so on.[7] It is clearly possible to inherit
a pattern of behavior; the animal is born with a memory of how
to do something that it has never done before. There are of
course many other examples (perhaps less striking ones) among
vertebrates; it is a well-known and accepted phenomenon. The
difficulties and the arguments usually revolve around how much
is learned and how much is inherited; there is even evidence that
to some extent an isolated individual can teach himself by trial
and error. But here we are not concerned with these fine points,
but only with the fact that in insects clearly the activities of the
neuromotor system are to a considerable degree transmitted
through the fertilized egg.

The big question is what is stored and how it is transmitted.
Undoubtedly it is not a whole elaborate set of instructions on
how to carry out a complex series of motor activities, but more
likely a simple set of individual responses to certain stimuli. In
an excellent discussion of the matter, E. O. Wilson[8] suggests that
the entire behavior of insects in a society might be broken down
into probabilities that any one caste (or age) of individuals will
respond in a set way to a variety of different stimuli. The unit
components, in this view, are restricted responses, which, for
each type of individual in the colony, have a different probability
of occurring. A good illustration of how group behavior results

from the individual responses can be seen in nest building. For instance, Sudd[9] showed in the weaver ant, which builds nests by folding leaves together, that if a group of ants start curling over a leaf, the individual that shows the most success (by having seized the end of the leaf) will soon be joined by all the others, and cooperative, group behavior will be the result (Fig. 10). All that need be inherited is the instinct to fold back a leaf and the instinct to work on that part of the leaf that already shows the biggest fold.

In these examples, what is inherited, and therefore stored, is a specific response. The great question is how genes affect the nervous system to produce the response; this is one of the most exciting areas of biology today. There is no doubt that genes do affect behavior, and that they do so by altering the responses. The evidence for this is enormous, not only in insects but in vertebrates as well. It is assumed that somehow the genetic information is pumped into the nervous system as it develops. There is a strong suspicion that it involves the arrangement of the neuronal network, but here we are walking on very thin ice.

One reason for the great fascination of the problem is that there are so many analogies between the structure of the genome

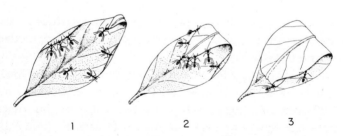

1   2   3

Fig. 10. The initiation of cooperative nest building in the weaver ant. When workers first attempt to fold a leaf (*left*) they spread over its surface and pull up on the edge wherever they can get a grip. One part (in this case the tip) is turned more easily than the others, and the initial success draws other workers who add their effort, abandoning the rest of the leaf margin (*center*). The result (*right*) is a rolled leaf of a kind frequently encountered in *Oecophylla* nests. [From J. H. Sudd, *Science Journal* (London, incorporating *Discovery;* 1963), Fig. 14–23.]

and the structure of the brain. They are both storage devices, and both, with the appropriate signals, will read off specific information, in one case specific protein, in the other a motor action. Both the brain and the genome have an extraordinary harmonious communication between parts: the modifier genes affect the action of other genes, and in ganglia there are modifying neurons that will enhance or inhibit the action of other motor neurons. One could continue with the analogy in some detail, but what we need now is hard facts that tell us more about the structure of the genome and of the central nervous systems. Then we shall have a clearer idea of how a behavioral response can be kept in the DNA of one cell and then later appear in a response of a complex multicellular organism to a specific stimulus. We shall return to the problem of the development of the nervous system in Part II of this book.

Despite this inheritance of specific responses, insects have a remarkable ability to learn, and they use this learning in their social behavior. Not only can ants learn to pass through mazes, but foraging ants and bees can learn and remember all the details of how to get to food and return home without error. They simply do not use these learning skills to imitate the actions of other, older insects.

The opposite is true in vertebrates, where much information is passed from one individual to another. A good example is the ability of titmice in Great Britain to open milk caps and drink the cream of bottles left at the door in the early morning. This skill has spread so rapidly to different parts of Britain that one must conclude it has been culturally transmitted. As one goes up the evolutionary scale the amount of information passed by learning from example, or by cultural tradition, progressively increases. There is probably a correlation between the fraction of the life cycle of an individual in which parental care is required and the degree to which the society depends on passing information by teaching-learning. It is certainly true that if one compares the child-raising period in the sequence rodents, lemurs, monkeys, primates, and man, there is a striking increase in the duration of the period of offspring dependency. The increase is associated with a progressive prolongation of development and a relative increase in the formative years, the period that is especially favorable for extensive learning.

Mechanisms of variation

We have divided the subject of inherited variation into (1) storage, that is, the mechanism of inheritance itself, (2) the mechanism of variation, and (3) gene action. This section on the second subject can be short, for it is the province of classical and modern genetics; the main relevant areas will be identified briefly.

The principal source of inherited variation in the genome is what used to be called a "point" mutation of a single gene. Now we know that a gene is a series of "codons," each of which is made up of three bases of the DNA. Certain combinations of these bases code for specific amino acids, and in this way the sequence of codons gives rise to a specific protein. It does not do it in one swoop, but each gene (that is, each sequence of codons) gives rise to a polypeptide chain, and subsequently these chains come together to form a protein. Sometimes the genes of the different chains are closely linked on the chromosome, but this is not always the case. Any error in the formation of any of the many codons that contribute to the formation of a protein will result in an alteration of one of the amino acids in the protein. The result may presumably be a change that in no way affects selection, a so-called "neutral" mutation. On the other hand, the alteration of one amino acid could have profound effects; an example is the change in shape of the haemoglobin molecule in the disease sickle-cell anaemia. Besides base substitutions, it is also possible to have insertions and deletions of bases which can cause mutation. In bacteria it is known that mutations can occur either as base changes in resting DNA or as errors during replication. However, the latter are largely corrected, for there is a special enzyme which removes sections that do not pair properly so they can pair again. In eukaryotes, besides the base alterations, there is the possibility that a whole section of chromosomes may be duplicated, deleted, or even inverted. Such chromosome aberrations play a significant but distinctly secondary role.

In sexual eukaryotic organisms, recombination through the segregation of whole chromosomes and crossing over within individual chromosomes is important in distributing the mutations in different combinations to new individual organisms. If particular combinations of genes give a phenotype of high selective value, that combination will be favored by selection. Therefore both mutation and recombination together are essential to make and distribute the variation effectively for selection. In asexual organisms recombination is lacking, and therefore its importance can be assessed by comparing asexual with sexual organisms. The rate of mutation may be higher in asexual forms, but the chance that three or four particular favorable mutant changes will accumulate in one individual is much less (that is, the time or the number of generations required is much greater). In asexual forms one clone will have to successively achieve the three or four mutants, while in sexual forms they may appear simultaneously in different individuals in a population, and could come together rapidly by mating and recombination. The whole subject of the mechanisms of handling variation in the genome is a vast one, which could easily occupy all the pages of this book.

There are also possibilities for variation in those parts of the storage system that lie entirely in the cytoplasm. It has already been pointed out that mitochondria and chloroplasts have their own DNA and RNA and synthesize the majority of their own proteins. This DNA is also capable of mutation and one can get altered organelles such as "petite" mitochondria or plastids without chlorophyll. The perfect model for this kind of change is the classical study of Sonneborn on kappa particles in *Paramaecium*, in which he showed that by growing the cells at different temperatures he could get the cytoplasmic particles to duplicate at different rates from the host cell. In this way he could increase or decrease the relative number of the particles per cell. This is a very immediate kind of selective process, and points up the fact that any change in a cytoplasmic particle is in direct competition with changes in the other particles within the cell. It must be tolerated or favored by its environment, which is the cell interior. It also shows that in their timing and frequency, organelle cycles can be quite independent of cell cycles.

There is another example of cytoplasmic variation that is directly inherited. In ciliates with well-organized rows of basal bodies or kineties, the rows are pinched in two each time the cell divides, and the rows in the daughter cells extend by the multiplication of the basal bodies. If there is an anomalous individual with fewer or more numerous kineties, Sonneborn[10] and others have shown that each time such an individual divides, the daughter cells directly inherit the anomaly. It is a direct transmission of cortical peculiarities as a result of the fact that the complex cortex in these organisms does not disappear at the point of minimum size, but remains intact and merely cleaves in two at each division.

We can add another mechanism of inherited variation that may play a significant role in development. In the most general sense it can be called a somatic mutation, a mutation that is not carried in the germ line. It could be a cytoplasmic change or a nuclear mutation. The idea that somatic nuclear changes might be important is an old one, but it is only recently that the concept has gained favor. Antibodies are produced in vast numbers by mammals, yet there are a few genes that determine their structure; the idea of one gene for one antibody must be rejected. An alternative is that the stem cells of the lymphocytes are capable of mutation, perhaps at a high rate, and those that produce appropriate antibody proteins are internally selected for within the organism.[11]

A change in the cytoplasm could also be of the genetic type and involve the DNA of one of the organelles, or it could be a chance variation. If a cell divides, the cytoplasm of the two daughter cells might differ significantly by accident, and such differences could be perpetuated in further divisions.

In some of these instances of somatic variation the individuals vary within a strict range, and I have called this "range variation." A good example is the size range of *Paramaecium* that was studied years ago by Jennings and his student Ackert.[12] In a clone the individuals will vary in length from approximately 90 to 160μ. If these two extremes are cloned, then each produces progeny with the same size range (90 to 160μ). Therefore the size of any individual is somatically determined, that is, determined by chance within strict limits. The entire range of sizes within which this chance variation can occur is inherited. This is clear from the fact that the largest and the smallest cells give rise to clones of identical range. The relevance of somatic variation to development in multicellular organisms and populations must come in later chapters; here I merely want to point out the fact that it exists.

Finally, there is another phenomenon that is well known in the development of populations. This is balanced genetic polymorphism, where in one population there may be two or more genotypes that exist simultaneously and bear some sort of fixed static or cyclical relation to one another. Well-known cases are the markings or the coloration of numerous insects. In each instance the complexities of the environment are such that the species is advantageously placed if it has two related genomes that coexist. In the case of butterflies, one presumes that the existence of different forms that mimic one another ensures that at any one time at least one of them will be safe from predators and able to mate. In other instances, where the change is cyclic, one form will have greater advantages than the other, depending on the season, and therefore the fact that the genome contains the information for both forms means that its chances of survival throughout the year are improved.

There is one interesting feature of the kind of polymorphism that is controlled by multiple alleles at different loci. If two forms exist, one predominating during one season of the year and the other at another, it is obvious that some mechanism is needed whereby the changes can be made quickly, but at the same time each character must be so placed that it will not be totally selected out at a particular season. In some beetles, for instance, which have different camouflage for protection from predators in the spring and the fall, the solution of the problem has been for the genes of the spring and the fall patterns to become linked so that they are genetically equivalent to one locus. By this amalgamation they become a kind of "supergene" and as a result they can be rapidly selected from one type to the other at the respective seasons, and furthermore by being linked they cannot be selected out completely and lost. Not only can genes affect or modify the expression of other genes; they can also be grouped in various ways, the balanced polymorphism of multiple alleles at different loci being an excellent example.

Since most multicellular organisms are derived from a clone, the cleavage of a diploid zygote, one presumes that such polymorphism does not play a role in multicellular development. Yet there is a close parallel. Note that these cases of seasonal polymorphism resemble what we have previously referred to as a temporal differentiation. The main difference is that in the seasonally balanced polymorphism the alternating genotypes arise by very rapid selection because of their supergene properties. If one compares this to an organism that undergoes a complete metamorphosis (say a fly or a sea urchin), then the larva appears to be one clone and the adult is derived from small buds (for example, imaginal discs) that give rise to the second clone. It would be absurd to think that the two clones (derived from either the zygote or the imaginal discs) have different genomes, but, as with polymorphism, different sets of genes are called forth at different times, in this case by rapid internal selection within the growing multicellular organism.

The amount of variation

It is important to remember that the amount of variation for any one organism is crucial and closely governed by selection. Because the subject is somewhat peripheral to our interest in development and because I have treated it in more detail elsewhere,[13] here I shall do no more than point out the main features of the subject. From this it will be seen that in the construction of the developing organism some aspects of the organism profoundly affect the amount of variation, and therefore are themselves under strict control by selection.

We begin with Darlington's[14] principle of compromise, which implies, among other things, that too little variation means excessively slow evolution, and too much variation means that all the gains would be obliterated by excessive shuffling of the genomes; there is an optimal amount of variation. If we turn to the factors that affect this variation, the first is the mutation rate. It is well known that the amount of mutation can be controlled; there are even mutator genes that control the rate of mutations on a chromosome.[15] The majority of the controls of variation affect the degree of recombination, and therefore the mere fact that an organism is asexual or sexual will play a role, because of the absence of recombination in asexual forms. The recombination can be controlled at the level of crossing over, for there is evidence of regions of chromosomes or whole chromosomes that are incapable of mutual exchange. The number of chromosomes itself affects the amount of whole-chromosome segregation or recombination, so chromosome number plays a significant part. The degree of outbreeding as compared with that of inbreeding in a population is exceedingly important, and this in turn is affected by many factors, such as the size of the population, or whether the organisms are unisexual (separate male or female individuals) or bisexual. Finally, there are more subtle factors such as the relative amounts of haploidy and diploidy in the life cycle; and the list could be greatly extended.

The fact that there are so many ways to achieve the same thing should come as no surprise. This is a perfect example of the very general property of natural selection that we have already stressed. Selection presses for a particular function, which in this case is the amount of variation, and the fact that it is achieved in many convergent ways is something one would have predicted. Notice, however, that in this particular case it is not always alternative ways that matter, but different mixtures of numerous possible ways. As long as the end result is optimal, the means of achieving it are of no consequence. What is measured is simply the sum total of variation, no matter how it is produced, no matter how many factors contribute.

Among organisms there are wide differences in the amount of variation. In some cases there is a virtual stop to all variation by closing the use of sex and reproducing entirely vegetatively. For instance, buffalo grass, which spreads across the Great Plains, is a fixed type that can exploit a particular environment with remarkable success. But if there were a new ice age, or a competitor that was even more efficient, buffalo grass would be incapable of changing. Most organisms go through periods in which they stop variation by vegetative growth, parthenogenesis, or some other form of asexual reproduction, but periodically (often seasonally) have a sexual cycle. In these cases one season is exploited by some rapid asexual reproduction, and after a sexual cycle the next season is faced with a new set of variants, some of which might be even better suited to the environment, perhaps in response to a change in climate or new competitors. By responding differently to different seasons this alteration of sexual and asexual cycles is functionally equivalent to balanced polymorphism, but as with our comparison with temporal differentiation the mechanism is again quite different.

Variation within the life cycle: gene action
Thus far all the examples we have discussed involve the entire life cycle, or, in the last case, even a succession of life cycles. If we now look at different parts of the life cycle, we see that certain stages will vary more than others when variation is present, and furthermore that some functions or activities of a developing organism will vary more or less than others that are occurring simultaneously. Let me give an example of the latter first.

Consider, for instance, the weight of a species of flying bird that has a particular kind of wing structure. Only a very narrow range of weights is possible, because in flying weight can vary only slightly without producing adverse effects. Too much weight would mean that the bird could never become or remain airborne; too little weight might produce aerodynamic problems of another sort that would result in a dangerous loss of flight control. It is not surprising, then, to find that young birds fly only when they approximate their adult weight, and that the various members of one species will vary little, compared, for instance, with terrestrial mammals. At the same time many other features of birds, such as the coloration of their eggs, might vary considerably.

We are even more concerned with the case where different stages of the life cycle differ in their degree of variability. To illustrate the case in the simplest possible way, the old rule or law of von Baer serves admirably: the young stages of two species resemble one another more precisely than the later or adult stages.[16] It is generally recognized that most variation occurs in the later stages of development; hence the large number of genes that are known for such things as bristle distribution and eye color in *Drosophila*, for example. The explanation routinely given is that variants occurring at an earlier stage embryonic development are most likely to be lethal, and the later a mutation appears in development the less the chance that this is so. Therefore there is selection for variations that occur at particular periods in the life cycle. One might add that if a viable mutant were to appear in early development, it could produce a major change in structure, entirely because of its position in the life cycle. As examples one could cite all the variations in gastrulation in animals, or the basic difference between the deuterostomia and the protostomia in the early evolution of invertebrates.

More important is the fact that the chance of having such a mutation is very slim, because of the high rate of appearance of lethals. But if it does occur and is selectively viable, it may have remarkable stability, remarkable lasting power. For, after all, it is still in a position where any further change is just as likely to be lethal. Therefore we find that periods in the life cycle of infrequent variation are periods which produce innovations that remain in circulation. This is another way of stating the simple fact that the ease of appearance of a variation in a population is likely to equal its ease of disappearance.

In those interesting organisms with a complex metamorphosis, there can be two periods of "youthfulness" and two periods of "maturity," all during a single life history from egg to adult. In the case of a housefly, the larva is a "mature" stage, as is the adult that arises from the small imaginal discs. It is not surprising, therefore, to find two peaks of high variation: one at the later larval period, and the other in the formation of the adult.[17] Furthermore, the early development of the egg and the early development of the imaginal discs are both periods of low variation. This illustrates the point that the degree of variability is determined by how much development must still take place. At any one moment, if complex developmental processes are still to come, then clearly any change is more likely to run into difficulties.

Thus far we have described major, improbable changes that

occur early in development. It is also possible to have rare mutations at late stages. Again we mean to imply by "rare" that they happen infrequently, and when they do occur they are of unusual stability. In this case their rarity is not the result of the danger that they might be lethal, but because they have achieved some remarkable feat whose chance of occurring had a very low probability. These are the cases where in one jump a new level of complexity is reached, and what was a late stage in the ancestor suddenly becomes a youthful stage in the descendants. In this way new properties emerge from the association or compounding of units on a previous level. In the biological world there are at least four such major steps or levels (they may, in fact, be more, but these four are obvious): (1) the origin of a prokaryotic cell (including the origin of life), (2) the origin of a eukaryotic cell, (3) the origin of multicellularity, and (4) the origin of social groupings. Let me hasten to say that, at least for the last two, there is every reason to believe that the innovation occurred more than once during the course of early history, but we are nevertheless dealing with relatively rare events.

For each of these levels it is possible to cite some evidence of stability. In the case of the origin of prokaryotes there is first the simple fact that the cell structures of all the prokaryotes that exist today have an amazing number of features in common: the structure of the cell membrane, the type of ribosomes, the simple construction of the DNA strands, and many more. On the chemical level, the same basic similarity is evident: the major metabolic pathways are basically the same, as are the majority of the enzymes that control the pathways. Perhaps the first and most fundamental point of stability is the genetic code and all the apparatus that it entails for transcribing and translating new proteins. The extraordinary thing is that this invention was so stable that, once achieved, it remained constant for both prokaryotes and eukaryotes. This means that not only did the triplets of bases in the DNA that specify certain amino acids remain constant, but also the *transfer* RNA's, so that the code can be translated to give the same amino acid in man as in a bacterium. There is, of course, much variation between species, but this variation

58

is small compared with the large number of structures and chemical constituents that are constant and ubiquitous.

On the second level, the formation of eukaryotes, the argument on structural stability of cell parts is equally applicable. Here there is the further intriguing possibility that the mitochondria and plastids might be symbionts. If one accepts this reasonable idea, it means that those symbionts must have virtually lost their own ability to vary and are now stable. One way this might have been achieved would be by the loss of any sexual activity or gene exchange. It is of course characteristic of parasites and symbionts to reduce radically their variance. They have an intricate close association with another organism, and any change is more likely to disturb than to improve that relation.

In the case of eukaryotic cells with plastids we have an innovation, a jump to another level, that makes for a radically different kind of energy processing. The fact that cells with and cells without plastids gave rise to wholly different lines of organisms in evolution is not surprising, considering the enormous functional difference between them.

If we turn to other cell structures, again we see evidence of remarkable stability. The structure and the distribution of the microtubules, the centrioles, the basal bodies, the flagella are similar in all eukaryotes. Although mitosis does show some variation in detail, the whole structure of chromosomes, their duplication, and their separation are also fundamentally similar for all eukaryotes. We can go on to the endoplasmic reticulum, the ribosomes, the nuclear membrane, the Golgi apparatus; all of these show the same kind of stability. The eukaryotic cell was a change, an invention of enormous success; it was so good, in fact, that very little modification was needed, or could be tolerated.

In the cases of the origins of multicellularity and social groupings in animals there is greater difficulty in citing this kind of evidence for stability. The evidence comes largely from seeing the different evolutionary lines and being able to trace them back to a common ancestor. For instance, primitive plant cells with plastids gave rise to filamentous cells with hard cell walls, and this structural condition is evident in the entire plant kingdom.

Similarly, in multicellular animal organisms there is invariably some kind of gastrulation, and despite its many modifications in detail in different phyla, there are aspects in common that are obvious in all animals.

Another property of all multicellular organisms is the communication between cells during development, just as on the level of social animals there is communication between individual organisms. But this is so general a property that it hardly serves the purpose of illustrating stability. Particularly in animal societies, it is more rewarding to show that certain social groupings, such as ant societies, have existed nonstop for many millions of years, and have proved their stability by showing little in the way of changes, yet they remain extremely widespread on the surface of the globe today.[18]

Superunits of gene action

We have pointed out instances of genes working in concert to produce supergenes, thereby making rapid selection easier and providing a mechanism for retaining those genes that may not be necessary at one moment, but will be essential the next. We also find that the actions of the genes may become bound together into "superunits." The advantage is that one such unit can be modified without adversely affecting other units; they are independent groups of interlocking gene actions.

The most obvious examples of such units involve the timing of major events in development. One developmental event could shift in time in relation to others, a phenomenon that de Beer[19] termed "heterochrony." For instance, the time of the appearance and maturation of the gonads occurs in some amphibians at the adult stage and in others at the larval stage (neoteny). Obviously, these two structures are entirely independent: the presence of one is not required for the other. The time at which they form can be genetically determined so that one can select for either precocious or late-appearing functional gonads. It is even possible in some forms that have an alternative asexual means of reproduction to completely eliminate all the sexual structures (as has occurred, for example, in the Fungi Imper-

60

fecti), yet the rest of development will be unaffected, for the building of the soma is not dependent upon the formation of sexual organs; they are developmentally separate.

Any new mutation that affected both the production of the sex cells and the development of the main body of the organism must have been selected against. We presume that it is of a distinct selective advantage to keep a number of the units of gene action of the organism quite independent of one another. The reason for this seems straightforward: mutations that affect a number of construction units are more likely to be lethal than those that affect only one. Or to put it another way, the fewer the interconnections of gene action produced by any mutant (that is, the less the pleiotropy), the greater the chances of its being a viable mutant. A viable mutant may be one that appears very late in development, such as the pigmentation of hair, eyes, or feathers, or one that acts in a small developmental unit that is independent of the others. The only major way in which these units seem to be affected by one another (a matter that we shall consider in more detail later) is in their relative timing. The genes that govern the heterochrony do not appear to seriously affect viability, but this, of course, is because the units have already become isolated by what has been called "internal selection."[20]

The idea that gene actions can be bound together in some way is a key component of the much-discussed recent hypothesis of Britten and Davidson.[21] They also see the need for grouping gene actions, but they suggest that this grouping takes place entirely in the genome. They postulate "integrator" genes that coordinate a number of "producer" genes; in this way the orders for close association are controlled by the DNA itself. This is an interesting suggestion, but one should not allow it to keep one from looking for other means of integrating gene actions, means that might well be largely cytoplasmic. There could, of course, be a variety of convergent mechanisms.

One of the important problems of the development of complex, multicellular organisms is how a variation system devised originally for unicellular organisms can manage all the genetic complications that arise. Some of those complications occur because the organism is large and lives for considerable lengths of time in a variable or seasonal environment. As we have seen, this may be helped by the clever device of supergenes that bind together the genes needed to produce balanced polymorphism. But a far more general device has been the formation of superunits of gene actions. By limiting the extent of the interlocking or pleiotropic effects of these actions it is possible to have variation apply to virtually all the structures that constitute the body of the animal or plant. Making the development of one organ independent of that of another has meant that the number of kinds of variation permissible is large, and therefore the possibility for evolutionary progress remains high. This subdivision of gene actions into superunits arose inevitably and naturally because of the internal selection within the developing organism; without it the world today might be populated solely with lower invertebrates and cryptogamic plants.

It is interesting that population geneticists are also finding that genes are often so closely linked that they do not act as independent alleles. In particular, Franklin and Lewontin[22] have suggested that this phenomenon is so widespread that the entire chromosome should be considered as a unit of selection rather than the individual alleles. As we have seen, there are apparently a variety of different ways in which the grouping of genetic information can occur: the grouping can occur entirely in the genome or in actions of the genomes.

Neutral variations

It should be added, as a corollary, that there is a considerable recent interest in a kind of biochemical variation that involves minor changes in the amino acid composition of proteins. Now that it is possible to carry out protein sequencing, it is evident that different populations and different species have characteristic deviations in the amino acid composition and sequence of many of their proteins. These are very small variations (as compared with the large ones we have been discussing) and it is not clear whether they represent neutral mutations, as some population geneticists contend, or have very low selective values, as is argued by others.[23] If we take the extreme position that they are neutral, it raises the interesting possibility that a significant change in protein structure could have arisen by pure chance rather than by selection. This, however, is neither a shocking nor a heretical view, and in fact it was clearly put forth by Sewell Wright many years ago in his concept of genetic drift for small populations. Small changes can indeed take place without selection and could ultimately lead to changes in structure of a significant sort, but eventually selection will take over and be the sole guide for any further change, be it viable variation or elimination, or for no further change at all, that is, stable equilibrium.

Summary

The contention has been that to understand development it is necessary to think of it in terms of variation, for what varies and what is subject to selection is the development. In order for variation to be of any value to selection, it must be inherited. In examining the mechanisms of this storage of information, it was stressed that both the DNA and all the rest of the cell components (which are DNA derived), including the active metabolism, constitute the sum total of storage from one generation to the next. The storage systems are all capable of variation, and we are especially clear on the details in the case of the variation of the genome, which is the province of classical genetics.

The amount of variation is controlled, and of particular interest are the differences in amount in different parts of the life cycle. Rare variations occur in early development, but also ones of special importance occasionally arise at the end of development, and begin new and higher levels in biological structure, such as the origin of prokaryotes, of eukaryotes, of multicellular organisms, and of social groupings of animals.

That variations are likely to be lethal in early parts of the life cycle emphasizes the fact that the process of development involves interconnections of gene-controlled steps, but in order for variation to occur, one part of the organism must be separate from the other, so that a change in any one part will not cause difficulties elsewhere. Therefore there has been a selective pressure for the grouping of developmental events into superunits of gene action.

We have dwelt on how developmental information is stored or inherited, and how it can simultaneously be capable of variation. Now we must examine how reproduction and variation arose in the first place (the origin of life) and how they became modified with each successive level of increased complexity.

5

The Levels of Complexity

We have stressed that there are a number of major steps in evolution where one leaps from one level of complexity to the next. These were probably rare, and in some cases they may be unique events. Here we shall examine these transition points in greater detail.

The origin of life

"Life" consists of so many elements that it is very difficult (and profitless to try) to define its origin precisely.[1] There are at least four questions that have concerned those interested in how life began on earth: (1) how molecules began to reproduce or replicate themselves, (2) how protein came to be synthesized, (3) how metabolism or continuous energy processing arose, and (4) how cells originated. I should like to discuss each of these briefly, and then concentrate on a fifth element which has not, to my knowledge, ever been systematically discussed: (5) how natural selection arose.

(1) The replication of molecules has, in the last few years, become the most actively discussed aspect of the origin of life. The elucidation of the structure of DNA by Watson and Crick made it possible to understand the mechanism of its replication, includ-

ing the remarkable fact that the copy is a negative one, a template of the original DNA strand. This, of course, is because of the specificity of the base pairing: an adenine will attach to a thymine, and a guanine to a cytosine. Furthermore, this system was also endowed with the property of varying (a mutation being a substitution of a base) and the inheritance of this variation by subsequent template replications.

Originally it was thought that only DNA possessed these properties, and that RNA was manufactured solely on a strand of the DNA. However, more recent work has shown that this is not the case: Spiegelman[2] and his co-workers have shown that RNA can replicate itself in a manner resembling DNA replication (and even more recently there is evidence that RNA can provide the mold for new DNA, a process promoted by the so-called reverse transcriptases).

In order to replicate RNA, an enzyme (a replicase) is needed. The obvious question is whether or not it would be possible to have replication without such an enzyme, because if we are to assume that the RNA was the first biological substance, then there could be no preexisting enzymes. This side of the question has been examined by Orgel[3] and his group. In the first place, the well-known experiments of Miller[4] showed that if one imitates primitive earth conditions, along with the energy available from electric storms, it is possible to produce a variety of organic molecules, including some amino acids, purines, and pyrimidines. Though Orgel's group is actively pursuing by experiment the question of how these bases might have become polymerized, they point out that there are sound theoretical grounds for proposing that primitive nucleic acid polymers might have been capable of catalyzing a replication reaction. It is also conceivable that there may be some inorganic catalysts which played a part. Enzymes are no more than especially efficient and specific catalysts; it is quite reasonable to assume that they might have arisen secondarily to supplement the less efficient primeval catalysts. (We shall examine how this might have occurred shortly.)

Spiegelman[5] has made use of his RNA replicase to examine the question of whether or not it would be possible to run selection experiments on RNA. He can, after all, make RNA in a test tube by adding various substances, including the bases and the replicase; now all that is needed is to select for some special property of this "naked gene." He has done a number of such experiments successfully by, for instance, selecting for an RNA that can reproduce efficiently with one of the components at a starvation level; in this way he produces a "nutritional mutant." In another type of experiment he selects for RNA's that form in greater amounts in the presence of some inhibitor, and again he finds mutants which breed true and produce RNA that is relatively resistant during its replication to the presence of the inhibitor. All of these experiments involve a series of transfers from one test tube to the next, each in the presence of the adverse condition. The growth rate of the RNA in each tube is measured and in this way it is possible to identify new fast-growing RNA's that may appear after a certain number of test-tube generations.

Spiegelman and his co-workers have done one selection experiment that is of particular significance: they have selected for fast replication under optimal conditions. After eight generations there was a sudden great increase in the rate of replication and they were able to show that this was due to a shortening of the RNA molecule. The RNA originally comes from an RNA virus and therefore it codes for a variety of proteins that have no bearing on the rate of replication. For instance, the protein for the coat of the virus and the attachment protein will no longer be needed in the test tube and therefore the portion of the RNA that codes for them could be eliminated and, as Spiegelman points out, they would then multiply faster simply because their RNA molecules would be shorter. Since more rapid replication or rate of reproduction ensures more offspring, one would assume that such a mutation would be of selective advantage, at least in the sequential test-tube experiments. However, they have lost their ability to infect host cells, indicating that they have indeed shed their "excess genetic baggage."

The important point is that although this would undoubtedly be true for artificial selection, for natural selection this experiment represents a rather crude evolution in reverse. This is an example at the very simplest level of something I plan to emphasize at all levels of complexity throughout this chapter. Though it is true that success in selection means success in producing more offspring or in making a greater contribution to future generations, it does not necessarily mean faster reproduction, even though that is the most obvious way of producing more offspring. Rather, it is important to produce offspring of such a nature that they in turn will be especially (in the statistical sense) likely to have offspring. We shall use the current phrase "cost of reproduction"; by this we mean that certain secondary nonreproductive activities involve a cost, but this cost is more than compensated for by the success in producing stable and successfully perpetuating offspring. In the case of the virus RNA, which is the reproduction unit in Spiegelman's example, only the replicase is needed for success in artificial test-tube generations. In nature, in order to succeed by selection, it must also develop other proteins such as coat or attachment proteins. Making these may slow the rate of reproduction because more RNA is involved, but the success of reproduction is ensured by paying this cost. Evolution by natural selection therefore means an increasing dependence on costly gene-controlled processes which in turn increase the probability of reproductive success.[6] Rapid multiplication alone is the most vulnerable method of reproduction.

It must be remembered that these experiments involve not only selection of the replicating RNA, but the combination of the RNA with a protein replicase. As was pointed out earlier, this does not represent a difficult theoretical problem because there may have been some other kind of catalyst when the first replicating molecules appeared. The advantage of the replicase is that one can do the experiments fairly rapidly, and the same advantage is conferred on the RNA in the virus as it infects its host in nature. Therefore it is possible to have some sort of selection take place on a "naked gene," but clearly such a gene alone has a very limited ability to evolve rapidly.

(2) The most obvious next step is for the molecule to manufacture its own catalysts so that it can lift itself by its boot straps. The step from replicating nucleic acid to making specific proteins

in addition is a huge one. (And note again that it must be considered as an increase in the cost of reproduction, but a very worthwhile increase if it makes possible specific replicases that will control the reproduction.) If we are considering protein synthesis in terms of how modern cells manage it, then there were undoubtedly a series of evolutionary steps, for the process is so complex it could hardly have arisen full blown. Consider all the parts of the process: *messenger* RNA is produced by replication on one of the strands of the DNA. There must be some mechanism that specifies which strand, and there is another to measure off what part of the strand. (It is possible that both strands are copied in RNA, and that the nonsense copy is destroyed.) The *messenger* RNA now reaches a ribosome, a complex structure made up of protein and *ribosomal* RNA (which has also been transcribed from another portion of the DNA). In addition, there are specific *transfer* RNA's, small molecules of some 80 bases (which also originated by transcription from the DNA), which are responsible for reading the code of the *messenger* RNA and placing the appropriate amino acids on a growing polypeptide chain. Each *transfer* RNA has 3 base pairs at the rounded end of its cloverleaf shape and these link to 3 base pairs on the *messenger* RNA. Since each *transfer* RNA carries its specific amino acid, by fitting into the successive 3 base regions on the *messenger* RNA as it passes through the ribosomes, the *transfer* RNA can (with the help of appropriate enzymes) hook together the amino acids into a polypeptide chain. In this way a polypeptide chain is produced in which the amino acid sequence exactly corresponds to the sequence of the triplet code on the *messenger* RNA.

Two related features of this remarkable process have been singled out as being especially primitive and therefore possibly representing the first events to appear in the origin of protein synthesis. One is the code itself, and the other is the *transfer* RNA. As Crick[7] has pointed out, every feature of the code points to a unique origin, unique in the sense that there have been no significant changes since its origin. If any change in the code itself had occurred, it would have wiped out all living forms by making nonsense of all the existing messages. This does not mean that all the codes for all the amino acids arose at one time; it is far more sensible to imagine this as a gradual process over many cycles. Once a code for an amino acid was established, an additional or redundant code (another combination of bases giving the same amino acid) would be possible. But it could not switch suddenly so that a particular triplet of bases now designated a different amino acid.

Transfer RNA is also thought to have a primitive origin, one closely linked with the code. Ohno[8] goes so far as to suggest that the invention of *transfer* RNA is the most appropriate event to take as the origin of life. The argument is necessarily much the same as with the code, since the *transfer* RNA's possess the codons for the universal code.

If RNA was the original replicating molecule, it was its own messenger, which removes one step in the complicated chain. One would then have to postulate that *transfer* RNA and *ribosomal* RNA, without any help of any protein, were able to manage the synthesis of the first proteins. As before, there is no theoretical reason why this could not have been so, but the argument is still thin and overly speculative. If one assumes that life began as bare nucleic acid, which subsequently became almost entirely a substance responsible for the production of protein, then one has to assume futher that RNA can do much of what the primitive protein could achieve (over and above its own special achievement of template replication). RNA must also have managed catalysis and some kind of primitive structure. The great advantage of of making protein is that it can do these things very much better.

Because of the fact that, once invented, proteins are manufactured directly by replicating nucleic acids, the possibility of evolutionary improvements (in the form of enzymes and structural proteins) by inherited variation and natural selection becomes enormous. It is not just the fact that sequences of amino acids have advantages for structure and catalysis; it is also that the possibility for variation is very much greater. Nucleic acids have only 4 bases, while proteins are made up of 20 amino acids (although no doubt the first proteins had fewer). So the code is not only a way to make proteins, but a way to produce great variation in phenotypic properties. The sequence of bases on a strand of RNA does not greatly alter its physical properties (although it might affect its postulated catalytic properties), but changes in the sequence of amino acids can produce spherical proteins, fibrous proteins, heat-stable or heat-labile proteins; the variety of protein properties is staggering, and all these differences can be coded in the relatively unexpressive parent nucleic acid. The invention of protein synthesis is the initial separation of the phenotype from the genotype.

(3) Assuming that the sequence we are using to describe the origin of life is correct, the next worthwhile investment (a further cost of reproduction) would be to invent a built-in motor, that is, to acquire metabolism. This would mean the synthesis of many more enzymes, and therefore the cost in nucleic acid production and protein synthesis would be large. It would mean that each cycle of replication would take much longer, as would the periods of translation. But if it were possible to provide a constant source of energy by chemosynthesis or photosynthesis, the effort expended upon all the new substances must have been small compared with the constant supply of energy that it provides.

In its most general sense, metabolism is a well-organized catalysis. Energy in the form of chemical energy or light is acquired and converted by the primitive organism to serve for its own chemical processes. This involves a whole series of reactions, each controlled by an enzyme, which are so arranged that the correct amount of energy is channeled not only for all the activities of the organism, but for all the thermodynamically improbable chemical reactions that need an energy push in order to take place. This means that by the strategic use of energy, reactions that would under normal conditions be unlikely or impossibly slow can now occur. Since enzymes control the rates of processes, each reaction is controlled so that not too much heat is generated, thereby preventing the organism from burning itself up. The new possibilities that metabolism brought to primeval organisms were staggering; it was a very great step forward.

Since harnessing energy involves a series of enzyme-controlled steps, it is rather easy to imagine a linear sequence of events that led from a very simple and unreliable energy processing to the complex one known today. A mutation that produces a new protein enzyme which in turn has special properties that promote some energetically useful reaction will clearly be maintained by strong positive selection. If now an additional enzyme can either modulate the first or add to the effectiveness of energy processing in some way, it will be equally desirable and therefore will be retained in the genome. Note that such steps involve an increase in the size of the genome to code for the new proteins; Ohno[9] provides a convincing argument that this may well have been by gene duplication. Some sections of the DNA were repeated by accident, in this way providing extra DNA. If now one copy remains the same while one mutates, it is possible to make two proteins where formerly only one was possible.

Today we no longer find organisms with a metabolic machine consisting merely of a few enzymes. Presumably the more complex later models were so much more successful that they outcompeted the early versions, which soon became extinct. (The same argument would apply to the hypothetical primitive RNA organisms, although some viruses, as we have seen, do use RNA as their genetic material.) The interesting thing is that just as all organisms seem to have the same replication system, the same code, and the same system of translation, they also have basically similar metabolic machinery. There are some variations, but they are small (and undoubtedly secondary) and this applies to all living organisms. So we may presume that life today had one origin for both its genetic and its metabolic systems.

(4) In our ignorance of molecular palaeontology of early organisms, it is impossible to separate the event of constructing a motor and the event of producing an outside membrane, that is, making the proto-organism into a genuine cell. The two events may have occurred at roughly the same time, or they may have been widely separated. The formation of a membrane involves protein, which may combine with other substances, especially lipids. The membrane has very special properties which are still not well understood: it can repair, expand, and contract with

magical ease. Perhaps the first membranes were not so versatile, but the ones that exist today seem all to be of a basically similar construction.

(5) We have briefly identified some of the major events that must have occurred in the origin of life; now we come to a question that is usually assumed rather than asked: when, in this chain of events, did natural selection take over and supplant accidental change?

Great stress has already been put on the fact that natural selection involves reproduction and inherited variation. This makes a very important assumption that does not apply to the origin of life. We assume that the reproduction is of animals or plants that are undergoing life cycles, for, as I have emphasized, it is the reproduction of the whole cycle that is important. This means, as I argued in Chapter 2, that there is selection not only for growth and size increase in the cycle, but for senescence and vulnerability to death by disease and accident as well. A built-in disposal system at the end of each life cycle is favored by natural selection, because, by kin selection, genes for the senescence of older individuals are carried by the offspring so that the younger individuals will have more of the limited resources available to them.

Coming back to the origin of life, we presume that the unit of selection is not an organism, but initially a nucleic acid molecule. Therefore one must postulate that in the beginning there were methods not only for constructing polynucleotides, but for taking them apart as well, to produce a cycle. It is a comparatively easy matter to visualize the death of a mammal, but the point is not quite so clear when we talk of the death of a polynucleotide. The separation of the bases and the sugars is really equivalent, but it is unclear that such a cleaving of the molecule arose as a result of selection during the earlier periods of the origin of life. The need for the degradation of the DNA is well illustrated in the case of Spiegelman's experiments on the selection of replicating RNA.[10] He has a built-in device to ensure the construction or growth of the RNA from its components, namely, the replicase. He also has a man-made method of terminating the life cycle of

the RNA; except for preserving enough to start the next generation, he throws the bulk of it down the sink. If he did not change
test tubes each generation, and thereby artificially create a cycle,
he would not be able to run his selection experiments; no more
than one generation per container is possible.

In discussing the origin of life we have stressed the fact that
the first catalysts could have been the polynucleotides themselves, or even some other substance found in the environment,
and that at a later stage the nucleic acids managed to synthesize
enzymes, which would be vastly superior catalysts. The implication is therefore that selection could have taken place on naked
genes, for they are theoretically capable of replication without
proteins. The process might be exceedingly slow and perhaps
not very regular, but it would occur. If catalysts are capable of
promoting RNA replication, they are equally likely to be able to
promote the depolymerization of the molecule. Or if it is not promoted catalytically, there may have been some other environmental characteristics on the early earth surface that would have
degraded the polymers. Once protein enzymes have appeared,
there are again two possible methods of elimination: by nucleases
or by some other environmental factor. As long as the break-
down follows the building up on some regular basis, natural
selection can occur; the RNA can grow by replication and the
old molecules are eliminated by degradation (reproduction of a
cycle in the most inclusive sense). Furthermore, the polynucleotides can vary by an alteration of a base and this change in information can be passed from one such cycle to another by the
replication process.

Our original question was: when did natural selection supplant
accident in the chain of events we have described for the origin
of life? Perhaps the best answer, in the light of what I have just
said, is that natural selection arose the moment a kind of molecule, by accident, became endowed with the properties of
variation and exact duplication, and with some means of regular
disintegration of most of its kind, leaving just enough to begin
a new cycle of duplication. Presumably this molecule was a
nucleic acid, and all the subsequent improvements, such as pro-

74

tein synthesis, acquiring a metabolic motor, and acquiring a cell membrane, appeared by chance and were retained through the aegis of natural selection.

One of the principal elements of my argument is that the evolution of recurrent life cycles has built up in such a way that all the information for any one life cycle does not come immediately from the DNA each generation, but an increasing amount becomes transcribed and translated during previous life cycles. Therefore there may be one or more generations between the manufacturing of a protein, or even the product of an enzyme-controlled reaction, and its moments of active participation in a specific life cycle. We are now in a position to look at how this arose during the first stages of the history of life.

The stage of "bare genes" (when nucleic acid simply produced nucleic acid) is the only stage in evolution at which all the information for each life cycle was read off at the beginning of each cycle. As soon as protein was manufactured, it was possible to make substances that could be carried over and used in a subsequent generation. As has already been stressed, the replicase is a perfect example: the replication of one life cycle is promoted by an enzyme made in the previous life cycle. The very moment such an enzyme was first invented it must have acted in this way, for transcription occurs not during replication but between replications.

Soon a few enzymes, along with structural proteins, must have managed an organized structure of some sort. It is hard to imagine what the first steps might have been, but as more and more proteins were synthesized, and more enzymes promoted different reactions, ultimately a simple prokaryotic cell was produced. At each stage of this accumulation, this building, we have a larger and larger collection of substances, now organized in an increasingly elaborate structure, which is carried over from one generation to the next. In a simple prokaryotic cell that goes through a series of divisions, many proteins and enzyme products are made long before they function, but we do not know when. Are all the replicases of one cell division made during the previous cell division, or are there some that go back numerous cell generations? Are the proteins in the cell membrane replaced rapidly, or do some of them persist over numerous cell cycles? It would be of equal interest to gain the same information from various eukaryotic organisms.

The important concept, however, is that we see the accumulation of all sorts of DNA-derived information that is ready for action at the beginning of a life cycle. This accumulation ultimately became stabilized at the level of a prokaryotic cell. The point of minimum size in the life cycle of *Escherichia coli* is the cell immediately following division and this small unit has a vast array of proteins of all sorts as well as innumerable substances that have been secondarily produced by enzymes.[11] DNA represents a very small amount of the bulk that bridges the generations, yet all the other substances were previously made directly by the DNA (all the proteins) or indirectly (all the other compounds), and they all, or at least a large share of them, make a direct contribution to the information needed to carry out the immediate life cycle successfully.

When looked at in this way, it is obvious that in all organisms that live today there is a cyclic component and also what at first appears to be a noncyclic one. The latter consists of the cell structure at the point of minimum size, while the former is the events such as growth that occur during the life cycle. But the elements that go into the stable structure of the cell each have their own cycle. All the parts within the cell are constantly duplicated or replaced: the molecules in the cell membrane, the ribosomes, in fact all the components of the cell. And this duplication and replacement occur without any loss or change in the basic structure of the cell. We know very little about the frequency of these alteration cycles; in some instances they undoubtedly correspond with the life cycle, and in others not. The whole process can be seen as the evolution of cycles within cycles. The "inner" cycles produce a stable (steady-state) end result: the minimum cell. The "outer" cycle brings the cell through the different morphological phases of the life cycle. As we have seen, the life cycle arose first; the inner cycles were added so that the life cycle could occur with even greater reliability, that is, greater reproductive success through natural selection.

76

Prokaryotic cells do not seem to have built-in self-destruct mechanisms; their elimination, which we showed was so important for selection, is apparently not achieved by senescence. The reason for this is undoubtedly that they, of all organisms, are especially vulnerable to the harshness and vagaries of the environment. Their numbers are kept in strict control by the limitation of available food and the fact that very small climatic changes will wipe out whole populations. Furthermore, bacteria, both in the soil and in an aquatic environment, have a large number of predators. Since, from an ecological point of view, they are colonizing species, that is, they depend heavily on dispersal and crash successes in pockets where the environment may be momentarily favorable, they perhaps do not need a senescence mechanism to limit their life span. This same argument may apply to many higher organisms as well, and we have already pointed out that the environment could be the factor that causes molecular disintegration or depolymerization. As one goes up the evolutionary scale, there is a progressive insulation of the organism from external fluctuations, and this decrease in harshness might explain the fact that senescence is particularly a character of higher organisms.

With the first appearance of the early prokaryotes there is a continued trend for selection to increase the cost of reproduction. Originally, rapid duplication was the prime criterion for reproductive success, but with the advent of a motor and all the structure that goes will cell duplication, synthesis, and metabolism, many other features automatically contribute to reproductive success. Obviously, those cells that find and process fuel most effectively will be more successful at reproduction. Any new quality the cell might acquire that will decrease its chances of destruction in the changing environment will increase its chances of success in producing offspring. Any new quality that simply gives the cell advantage in competition with other cells for food or for shelter will be favored by selection, and will eventually play a far more significant role than the speed of making copies.

It is here, in fact, that we see the advantages of size appear for the first time. With an increase in size and the concomitant increase in complexity, specialized structures appear. For instance,

77

flagella arise which permit the cell to move; it can go toward food or away from adverse conditions. Spore formation appears which protects the cell against drought and other adverse environmental changes. Any increase in size and complexity will automatically mean a decrease in the rate of duplication or reproduction, but what is lost in duplication speed is more than gained by successfully competing for limited resources and being less susceptible to environmental hazards.

Not all gains need necessarily be related to increase in size, a point emphasized before. A good example is found among present-day prokaryotes. As Pardee[12] has emphasized, there are some bacteria with very restricted diets that are capable of rapid growth and have short generation times. There are others with a totally different strategy: they can eat a wide variety of food substances, but because of this ability they must, at least potentially, be able to have a large number of different enzymes to handle such an array of substrates. However, they pay the price by reproducing very slowly. Obviously, in some environments that exist today, such complex, multienzymed forms have an advantage over the specialist that can utilize only one substrate, even though the latter reproduces much faster than the former. One species counts on one specific substance being present and then exploits it by speed of reproduction; the other can manage a large number of substrates and exploits this variety where the specialist would fail.

If, therefore, one looks at the evolution of natural selection, one must conclude that at first there was a greater emphasis on the speed of reproduction alone; it was the era of catalysis. But soon speed was replaced by other, more effective means of competition, which involved an increase in complexity, an increase in the cost of reproduction, and therefore a wide variety of new and more important avenues of reproductive success.

The origin of eukaryotes

One of the most conspicuous differences between eukaryotes and prokaryotes is size. The accepted view is that some of the cell organelles, particularly plastids and mitochondria, were originally prokaryotic symbionts. There are also many other structural advances, such as the nuclear membrane, the centrioles, the spindle, the Golgi apparatus. All of this means that at the point of minimum size in a life cycle a eukaryote carries much more information. It not only carries more because it has more DNA, but it also has far more previously made DNA products. It is particularly interesting that plastids and mitochondria have their own DNA, as one would expect from their ancestry as symbionts. They are no longer completely autonomous and are held under control by nuclear genes of the host cell, yet nevertheless each of these organelles goes through its own "life cycle." Put in this way, we see that eukaryotes have a far greater number of cycles within cycles than prokaryotes. Furthermore, there are structures within structures (what Bernal[13] called circles within circles) and the total number of substances and structures that produce the "inner" cycles is relatively enormous in eukaryotes. The point of minimum size in the life cycle has become large and complex.

What were the selective forces that might have led to the origin of the eukaryotic cell? One assumes that there was a selective advantage for an increase in size and complexity. Certainly if one looks for variety of form, or of specialized adaptation, eukaryotic cells are far more diverse than prokaryotes. Yet, at the same time, eukaryotes were not successful at the expense of prokaryotes, for the latter not only have existed on earth longer, but remain by far the more abundant in terms of numbers of individuals.

Eukaryotic cells are typically on the order of $10\times$ larger ($1000\times$ greater volume) than prokaryotes, and this considerable difference has many ecological consequences. The most obvious is that eukaryotic cells can eat bacteria, but the reverse is not true (except by parasitism); this is an important selective difference. Certain types of eukaryotic cells, such as free-living ameobae, have become predators, and the small bacteria are their prey. In the same vein, the size increase confers a distinct advantage in terms of rate of movement. It is generally true for swimmers that the speed of the swiftest among all species of a particular size is directly proportional to its length (Fig. 5). This being the case, the fastest eukaryotic cells can move on the average $10\times$ faster than the most rapid prokaryotes.

Perhaps an equally important, but quite different, virtue of the eukaryotic cell is that it is a superior building block for multi-cellular organisms, a matter to be considered in detail presently. Here let me merely point out that from an engineering point of view its building accomplishments are quite fantastic when one compares them with those of prokaryotes. Even in the unicellular protozoa and algae, the variety of form, compared with prokaryotes, is enormous. Again we can assume that the reasons for this are directly related to size: the efficiency of the motor, the capabilities of the storage system, and those of the variation system are all bound to be size-limited. A eukaryotic cell can do many things that a prokaryotic cell cannot do, and for those that both can do, the eukaryote often manages better. Eukaryotic cells fall into two distinct types: those with and those without hard cell walls. This is a very important dichotomy because the presence or absence of the wall has a profound effect on the mode of construction (development) of the organism and on its locomotory physiology.

The increased size of the eukaryotic cell has permitted it two separate but closely related revolutions. One is ecological: the larger cell has been able to find new niches that did not exist for its smaller progenitors. The other is structural: the larger cell can engage in a wide variety of cell differentiations that are apparently denied to prokaryotes. In particular, it is a unit of construction especially well adapted to the mechanical needs of multicellular life cycles.

The origin of multicellular organisms

The step from unicellular to multicellular organisms undoubtedly occurred a number of times; it is certainly not a unique event. The evidence comes largely from the array of colonial forms that exist today, many of which are clearly unrelated to one another and must have always been independent (Fig. 11). Before giving specific examples I should mention that I am using the word "multicellular" here in the most general sense, to include all those forms that have some stage in their life history in which a number of cells are confined or attached together.

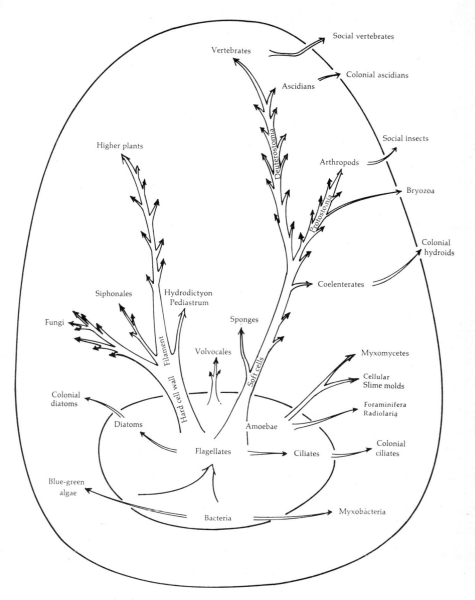

Fig. 11. A hypothetical scheme representing some of the principal independent attempts at producing multicellular, colonial, and social organisms. The innermost ellipse contains the one-cell forms, the large ovoid contains the multicellular forms, and the colonial and social forms are outside this ovoid.

81

Defined this way, bacteria provide an excellent example of an independent step toward multicellularity. During their vegetative stage the myxobacteria have motile, rod-shaped cells that move together in a swarm. These swarms wander as they absorb nutrients, and once the food supply is depleted they accumulate at central collection points and form multicellular fruiting bodies. In *Chondromyces* these are stalked and the cysts contain many bacterial cells, which burst out upon germination (Fig. 12).

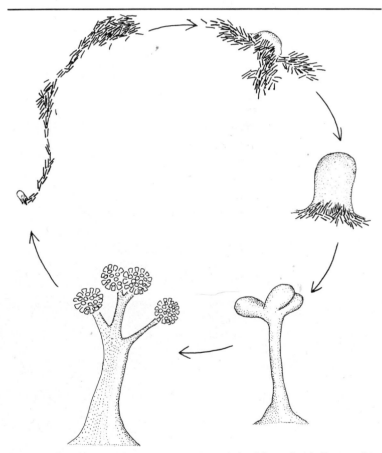

Fig. 12. Semidiagrammatic representation of the life cycle of the myxobacterium, *Chondromyces crocatus*. A mature fruiting body (*lower left*) bears cysts, each of which liberates numerous bacterial rods, which accumulate in progressively larger groups, ultimately producing a new fruiting body.

The blue-green algae undoubtedly had a separate origin for their multicellularity. These large, photosynthetic, prokaryote-like algae were extraordinarily successful as the principal flora in Cambrian times, and even today they are exceptionally adept at growing under a large variety of difficult conditions.

The filamentous eukaryotic plants may have had a single or a multiple origin. In external structure they are strikingly convergent with blue-green algae. They presumably gave rise to the fungi, numerous groups of algae, and all the higher plants.

There are other algae, such as the red algae or the Chlorococcales (for example, *Hydrodictyon* and *Pediastrum*), that have a number of unique features and therefore may have arisen independently.

Volvox and its relatives are of particular interest because, even though they are green algae, they show close similarities with the sponges, in that both have an inversion stage in which a single layer of cells turns inside out. It is presumed that sponges are close to the beginning of all multicellular animals; at least they are one of the most primitive forms that exists today.

There are numerous multicellular amoebae: the foraminifera, the radiolaria, the myxomycetes, and the cellular slime molds. The first three groups have syncytial stages during most of their life history and move from the multinucleate condition to a multicellular one only during restricted moments. In the cellular slime molds there is a true aggregation of cells to form fruiting bodies, similar (convergent) to what we describe in the myxobacteria (Fig. 13).

Finally, there are curious forms, clearly of independent origin, to be found among the so-called colonial ciliates and the colonial diatoms. These must be independent in their origin because their building blocks are totally different and are not closely related in ancestry.

It would be difficult, with such an array of possibilities, to examine each of these instances of beginning multicellularity in any detail without finding the successive descriptions bewildering. Instead my plan will be to discuss separately the origin of multicellular organisms in a terrestrial environment, and then do the same for an aquatic environment. For both conditions the main theme will be to attempt to understand what aspects of the many-celled conditions are adaptive; why did selection favor the stick ing together of a number of single-celled organisms to produce multicellularity? Again there is possibly a general advantage to

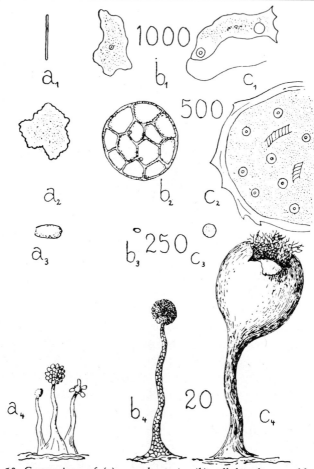

Fig. 13. Comparison of (*a*) myxobacteria, (*b*) cellular slime molds, and (*c*) myxomycetes: (*top*) the individual cells of each type of organism; (*upper middle*) cross section of their fruiting bodies; (*lower middle*) fruit cysts or spores; (*bottom*) their mature fruiting bodies. [From E. Jahn, *Die Polyangi-den* (Borntraeger, Leipzig, 1924).]

size, but we need to be more specific and to understand with greater precision the factors involved. We must remember that success in selection means success in reproduction. This success depends heavily on finding the most efficient and effective ways of turning energy into making copies. Therefore one of the first things we must look at is whether or not a particular multicellular form has some advantage in feeding efficiency; whether or not it can get food energy in amounts and in kinds that are unavailable to the single-celled form. Secondly, we must look for other ways in which this process might be helped. For instance, if an organism devised an efficient way of distributing itself over wide areas, it would be more successful in making copies simply because it would find more food patches. Or an organism may be helped because it can move in such a way as to find food more effectively and therefore have more offspring. Even the presence of resistant stages and complex life cycles that permit survival in a fluctuating environment will mean that more offspring will ultimately be produced, because the organism's competitors literally did not live through the difficult periods.

For the most part, terrestrial unicellular organisms are non-photosynthetic. There are a few exceptions, such as *Protococcus*, a green alga that grows on the north side of trees in temperate climates, and red snow, which is a unicellular red alga. Most of the nonphotosynthetic forms live in the top layer of the soil or humus. Here we find a myriad of flagellates, ciliates, many species of amoebae (and, of course, large quantities of bacteria). None of these forms can exist in their feeding or vegetative state in dry soil; to grow they all need a considerable amount of moisture between the grains of dirt. At first it would seem that this fact blurs the distinction between terrestrial and aquatic environments, but, as we shall see, the division line is reasonably sharp. Single cells would shrivel up and die, or at least would suspend all activity, without moisture, simply because they do not have the impermeable coats that are found in larger organisms. The only cells that are protected are the spore or cyst stages, which are inactive. The soil is perfectly designed, with its many minute capillary interstices, to retain large quantities of moisture, even over periods of considerable drought, providing a perfect moist chamber for unicellular growth and movement.

One of the principal characteristics of the soil (and in this respect it differs in degree from an aquatic environment) is that the food is likely to be unevenly distributed. It is organic matter which can be eaten by bacteria, or bacteria which can be eaten by amoebae; in both cases the food will most likely be patchy in its distribution. A drop of organic matter will land in one spot and before long that spot will be a mass of bacteria that can consume it; in turn those bacteria will soon be surrounded by predatory amoebae.

How do the bacteria arrive at the spot in the first place? There are two methods: one is simply by growth. If a bacterial cell happens to be in that locality when the drop of organic matter appears, then it can quickly take in the food and multiply rapidly. If a cell cycle or a generation is a matter of an hour or less, in a very short period there will be an enormous number of cells feeding at the spot. (Let me point out, in anticipation of the discussion to come shortly, that the presence of a bacterial cell in that spot in the first place depends on its having an effective dispersal system.)

The second way in which bacteria can reach the spot is by moving toward it. The usual method of doing this is by flagellar motion, something that the minute bacterial cells can manage in the thin films of water that lie over the surfaces of the soil particles. The cells could be moving at random (and no doubt many do exactly that) and by chance bump into the spot. This is again a means of effective dispersal. Much more interesting is the fact that bacteria can orient chemotactically to specific substances. This has been studied in some detail recently by Adler,[14] who finds that a cell of *Escherichia coli* moves in a straight line, stops, turns slightly, and goes in another straight line. If there is a gradient of an attractant substance, the cell will go farther when the concentration of that substance increases rather than decreases. The important contribution of Adler is to show that there are specific binding sites on the surface of the cell for the attractant.

Furthermore, the attractant need not necessarily be metabolized. One presumes that originally such chemotaxis was positive, toward substances that could be eaten, but Adler produced mutants that failed to metabolize a particular sugar, yet continued to be attracted to it in a gradient.

We can take this same story on to the next level. It has been known for a very long time that both mammalian leukocytes and soil amoebae are attracted to food, in this case their prey being bacteria. It is possible to produce gradients of bacterial extract and show that the amoebae will move up the gradient. With the help of bioassays we now know at least two of the substances that are given off by bacteria which must play a part in this food-seeking chemotaxis.[15] They are cyclic AMP and folic acid.[16] Both these substances are small molecules (hence rapid diffusers) that are secreted extracellularly by the bacteria. Both molecules are heat stable and subject to enzymatic degradation. Less is known about the elimination of folic acid by the amoebae, but with cyclic AMP we know that the soil amoebae produce an extracellular phosphodiesterase which specifically converts cyclic AMP to $5'-AMP$, a substance that is chemotactically inactive. It is obvious that such an enzyme helps maintain a steep gradient of attractant, thereby improving the efficiency of the food-finding chemotaxis.

There is one further fact of interest concerning soil-amoebae chemotaxis that bears on dispersion mechanisms. Amoebae apparently give off some substance to which they are negatively chemotactic,[17] that is, they repel one another, and move as far apart as possible. Thus in a region devoid of bacteria they disperse more effectively in the search for food. In a uniform patch of bacteria, the same mechanism permits more effective grazing in the presence of a uniform distribution of food.

Besides a patchy spatial distribution of food, the environment can be patchy in a temporal fashion by changes in the environment. Since patchiness in general seems to stimulate complex adaptations, it is not surprising to find that both single-cell prokaryotes and eukaryotes respond by producing a complex life cycle in which a resistant spore or cyst alternates with vegetative

87

or feeding stages. A decrease in humidity in some cases, and invariably a disappearance of food (that is, starvation), will produce the resistant stage. The organism is preparing for adverse, nongrowing conditions. These conditions may be prolonged and severe; therefore the chance of surviving them will be enhanced if the cell goes into a resistant storage phase. The selection pressure in a fluctuating environment will favor such a two-stage life cycle, this temporal differentiation within the life cycle.

Thus far we have been laying the groundwork for an understanding of how multicellular organisms might have come into being in a soil environment. If bacterial cells, or amoebae, are to stick together and form a large group, what is the selective advantage; what could have favored such an association? Obviously a larger organism would be produced, but why would such an association have a competitive advantage over the ancestral unicellular forms? There are two conceivable advantages, and here we shall examine the evidence for each. One is an advantage in feeding; the other is an advantage in dispersal.

The advantage of increased size for feeding is easy to see for mammalian carnivores that must pull down their prey, but it is much less obvious in the case of soil microorganisms. The myxobacteria produce an excellent example, as Dworkin[18] points out (Fig. 12). In these organisms the cells are at all times gregarious, a fact that can be shown in a number of different ways: when a cyst germinates, the bacterial rods all stay together, and never disperse. Furthermore, if two such groups of gliding, moving rods meet they fuse to form a larger mass of swarming cells. Even the locomotion is stimulated by a multicellular condition, for a mutant is known that moves only in the presence of other cells. In those species that form myxospores (where a single bacterial rod rounds up encysts) under certain nutrient-poor conditions, a high density of such myxospores is necessary in order to have them germinate. (It is possible to wash the spores and obtain a substance that will stimulate germination in isolated spores.) Everything about their vegetative existence points to the fact that they are happier in groups. The reason for this is likely to be found in the kind of food they eat. They do not thrive solely on small organic molecules, but mainly on insoluble, particulate detritus such as starch, the carcasses of bacteria which they lyse, protein, or cellulose. The suggestion of Dworkin is simply that in order to break down these difficult foods they need to attack them with massive doses of extracellular enzymes, and the more cells there are together, the greater the chance of their being able to produce sufficient enzyme to do this effectively. The advantage of size, according to this hypothesis, is the massive barrage of enzymes that many cells provide to break down recalcitrant substrates.

The amoeboid organisms provide a similar case. These are the myxomycetes, which produce a large plasmodium or multinu-

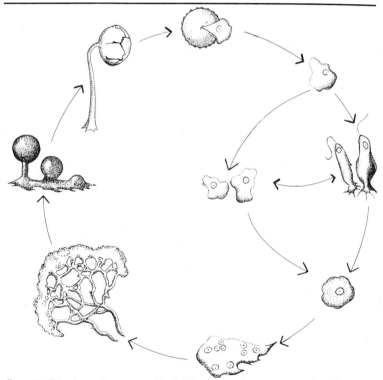

Fig. 14. Diagram of a generalized life cycle of a myxomycete. The spore germinates, giving rise to a cell which, depending upon the environmental conditions, is either a myxamoeba (for example, in a dry environment) or a flagellated swarm cell (for example, in a wet environment). After fertilization, the zygote grows into a large multinucleate plasmodium that eventually turns into many spore-bearing fruiting bodies. [Redrawn from C. J. Alexopoulos and J. Koevenig, *Slime Molds and Research* (Biological Sciences Curriculum Study pamphlet No. 13; Heath, Lexington, Mass., 1964).]

cleolate cell mass (Fig. 14). The life cycle is sexual. The spores will liberate haploid gametes which, either as amoebae or by transformation into flagellated swarmers, fuse to form a diploid zygote amoeba. This zygote now grows by repeated, synchronous nuclear divisions (and occasionally neighboring plasmodia will fuse) so that eventually one very large sheet of multinucleate protoplasm covers a large area of soil or humus. Myxomycetes are similar to myxobacteria in that they also are able to attack macromolecular and particulate food, and this they can apparently achieve with remarkable efficiency. As effective consumers of this difficult food, they outconsume their competitors and thereby have a selective advantage. The plasmodium moves; the protoplasm pulses back and forth in the blood-vessel-like channels it forms, but there will be more net flow in one direction than another; like a series of waves on an incoming tide, the advancing edge will slowly move forward. Furthermore, it will move toward food; it is capable of chemotactic orientation. In particular, it is sensitive to sugars and amino acids and will move up gradients of these substances.[19] Again, with the advent of food depletion or starvation, the protoplasm forms a series of small bumps, each of which (in certain species) will rise into the air and produce small stalked fruiting bodies. The stalk itself is a noncellular deposit; the sporangium is cut up into a series of uninucleate compartments as meiosis proceeds, so that ultimately the sporangium contains a mass of resistant, haploid spores. In this case, as in the myxobacteria, it appears very likely that, by accumulating large masses of feeding protoplasm, the organism can obtain certain types of relatively indigestible food more effectively than unicellular forms. Because of their size they occupy slightly different environmental niches in the soil from the smaller unicellular forms, and for this reason both the larger and the small forms can comfortably coexist.

This brings us directly to the second advantage of increased size, namely, dispersal, which is another way of successfully competing by finding more food. The best way to illustrate the point is to compare the life cycle of the cellular slime molds or acrasiales with that of the myxomycetes (Fig. 15). During the feeding

90

stage the amoebae of *Dictyostelium*, unlike the myxomycetes, remain entirely separate and unattached. Each small cell engulfs bacteria by phagocytosis, and periodically divides by binary fission; each daughter cell wanders off to graze independently. In this case, clearly, feeding is not done as a cooperative multicellular venture, in contrast to the two previous examples, but is strictly a unicellular effort. When the food is gone and the starvation reaction sets in, the amoebae swarm in to central collection points, a chemotactic process governed in some species by the secretion of cyclic AMP and a specific phosphodiesterase (which are also known as acrasin and acrasinase). The resulting cell mass forms a slug that moves as a body and orients toward light and heat. After a period of migration it rights itself and rises into the air, to form a small fruiting body. In this case the anterior cells turn into the stalk, the outer casing of which they first secrete, and then they enter into the top of the cylindrical casing by a reverse fountain movement, swell by becoming filled with a vacuole, and die. Their carcasses have rigid cellulose walls which provide cross struts that must help in supporting the delicate stalk. The posterior cells turn into spores, each amoeba becoming condensed into an elliptical spore case.

In this organism there is no multicellular feeding, yet a multicellular spore-bearing body is formed. Furthermore, it is formed at a considerable reproductive cost, for in each fruiting body a portion of its cells die in the process of stalk formation. From the point of view of selection, there must be a considerable advantage to such a fruiting body if it is going to exist at the ex-

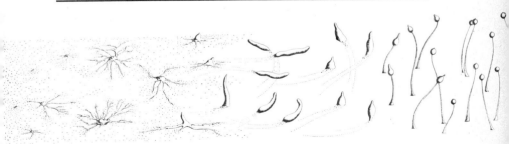

Fig. 15. The development of *Dictyostelium discoideum* from the vegetative stage (*left*), through aggregation, migration, and the final culmination stage. [Drawing by Patricia Collins from J. T. Bonner, *Scientific American* (June 1969), pp. 82–83.]

pense of losing a number of its cells. One can only conclude that a fruiting body has a large adaptive value in its ability to disperse spores. Why this should be the case is less clear. The fruiting bodies will form in small cavities near the surface of the soil. Perhaps this is the region where numerous worms and arthropods crisscross, and it is true that the spores from a spore mass will cling to anything that touches it. However, no experiments have been done, for instance, on the relation of fruiting-body size to dispersal efficiency; there is an important opportunity here to understand the mechanics of this dispersal.

There is considerably more evidence that selection pressure for fruiting bodies exists and is a widespread phenomenon. We already have seen such evidence in the stalked forms of myxobacteria and myxomycetes; it is an equally obvious and common phenomenon among the vast array of terrestrial molds. The fungi have separate hyphae which invade the soil by spreading as far apart as possible from one another. In fact, there is clear evidence that the growing tips are repelled by neighboring hyphae, which means that, like the solitary-feeding cellular slime-

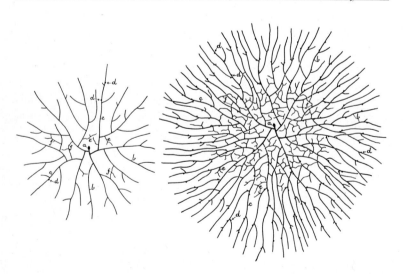

Fig. 16. Two stages in the growth of a mycelium of the basidiomycete *Coprinus* from a spore (*a*). [From A. R. H. Buller, *Researches on Fungi* (Longmans Green, London, 1931), vol. 4, Figs. 87 and 88.]

mold amoebae, they are grazing over a nutrient-rich area as evenly as possible (Fig. 16). It means that, even though the hyphae are attached, each tip is feeding as though it were a separate organism. It is, of course, well known that the growing hyphae will orient chemotactically toward the food. D. R. Stadler[20] showed that in the bread mold *Rhizopus* this was due to the fact that certain foods produced a substance that inactivated the substance responsible for the hyphae growing away from one another. In this way the negative chemotropism of the repellent is removed near the food and as a result of the steepened gradient the hyphae will orient chemotactically toward the food. Although this evidence supports the view that, as far as feeding is concerned, these molds are not unlike solitary soil amoebae, they resemble the acrasiales in that they possess numerous stalked fruiting bodies containing many spores. In this case, the protoplasm within the hyphae flows to central collection points, and builds a sporangiophore that terminates in a mass of spores; there is no new growth in this process, but an aggregation of the protoplasm that has accumulated in the vegetative hyphae. Again it is presumably stimulated by starvation, although the food depletion may occur just a short distance behind the advancing vegetative hyphal point. In this case, too, one is forced to conclude that the sporangium, which is characteristic of a vast number of species of soil molds, must be an adaptation for effective dispersal.

In fungi, among both the ascomycetes and the basidiomycetes, there are many examples of compound fruiting bodies. In these instances, at the moment when fruiting-body formation begins, there is a sudden switch, and instead of the hyphae growing away from each other, the protoplasm flows centripetally from all the vegetative hyphae and accumulates at specific points. The new aerial hyphae tend to adhere to one another and grow in close association (Fig. 17). The result is a larger spore-bearing body which is made up of masses of hyphae. The higher basidiomycetes produce mushrooms, which may achieve considerable size and contain vast quantities of spores. Again there would seem to be an important need for a study of the relation between fruiting-body size and spore-dispersal strategies.

In all these instances of fruiting-body formation there is evidence of tropisms, a fact which undoubtedly reflects that it is important both to have a receptacle for a large number of spores and to place it strategically for effective dispersal. In the first place, light orients many of the fruiting bodies (and in some cases it is also responsible for initiating fruiting-body formation). So we presume that to be near the surface is often an important factor. Light tropisms are well known among mushrooms, many simple molds (for example, *Phycomyces*, which has been studied extensively), the cellular slime molds, and even the myxobacteria.[21] In the cellular slime molds the ability to orient toward light was examined in a whole series of fruiting bodies of different sizes.[22] Within limits, the larger the fruiting body, the more responsive it was to unilateral light. This suggests the possibility that in these organisms, if orientation to light is selectively advantageous for dispersal, a certain size is mandatory in order to achieve the necessary light response.

Besides light there are other orientation mechanisms. In mushrooms, gravity plays an important role; it is the main factor that produces favorable orientation for spore dispersal.[23] In cellular slime molds, which are too small to be affected to a significant degree by gravity, there is evidence that the rising fruiting bodies give off a volatile substance that causes a negative chemotaxis; if two fruiting bodies are rising in close proximity to each other, they will repel each other.[24] It was also shown that this substance was responsible for the orientation of a solitary fruiting body. Since the gas was produced equally on all sides, it caused the fruiting body to rise at right angles from the substratum. It is unknown whether or not this gas-induced orientation is a general phenomenon, although there are suggestions that it might occur among the smaller fungi.[25] By such methods it is clear that not only will there be a specific orientation with respect to the substratum, but also a group of fruiting bodies will stand equidistant from one another. All this, we presume, is favored by selection in the struggle for the more effective dispersal of spores.

To return to temporal changes and their effect on life histories, it should be pointed out that even primitive multicellular organisms with complex life cycles, such as cellular slime molds, can have further complexities as a response to changing environments. Some strains of some species have alternative developmental pathways and the environmental conditions determine which pathway the organism takes. If the conditions are very crowded and the amoeba population is overly dense owing to an abnormally high concentration of food, some forms will form microcysts. The isolated amoebae, before aggregation, become encysted, much like solitary soil amoebae. This is a viable resistant stage that later can be induced to germinate. If conditions are very wet, in fact positively submerged, some strains of some species will form macrocysts. In this instance aggregation occurs, but instead of the aggregate forming a fruiting body, curious large cysts are formed (which recent work suggests are zygotes). Obviously, in exposed soil one can expect such environmental variations, and certain populations of slime mold are specially adapted to meet these different contingencies. In this case the adaptation to the temporally patchy environment is by alternative pathways of development rather than by a sequence of fixed stages.

In the soil there is a clear indication that multicellularity arose for two reasons: (1) in a few instances to develop a new kind of mass feeding which is successful largely by avoiding competition with smaller unicellular organisms, and (2) to develop more effective dispersal methods. These generalizations apply to a number of lower organisms, including the fungi. Since from classical phylogenetic studies we presume the fungi to be polyphyletic, and since there are aquatic fungi, it is not clear whether or not the terrestrial examples came from aquatic forms and did not attain their multicellular condition as terrestrial plants. Both processes could have occurred in fact. Once the fungi were established in the soil, they followed much the same pattern of food seeking and spore dispersion found in totally unrelated groups —amoebae and myxobacteria.

If we turn to a consideration of the origin of multicellularity in an aquatic environment, we find two contrasting conditions: one is existence on the bottom and presents almost identical pro-

Fig. 17. Stages in the development of the fruiting body of the basidiomycete *Pterula gracilis*. The fruiting body began as a single hypha, and eventually the spore-bearing hyphae jut off at right angles to the main axis (Redrawn from E. J. H. Corner).

blems to that of the soil; the other is the existence of free-swimming organisms.

As in the case of soil, those aquatic forms that live on the bottom can be patchy and all the arguments concerning the distribution of bacteria and unicellular protozoa are the same. It is not surprising, therefore, that the morphology and the distribution of water molds are similar to those of terrestrial molds. Vegetative hyphae penetrate into the substratum, and once they have exhausted the food they send out sporangia into the water, and motile zoospores are liberated (Fig. 18). So here there would seem to be a similar argument for dispersal; whether or not these sessile forms benefit by feeding in large masses is doubtful. It might be possible to argue that mass feeding is advantageous in the curious colonial *Labyrinthula* (Fig. 19), which feeds in a large net, and has an expanding edge not unlike the myxobacteria or the myxomycetes.

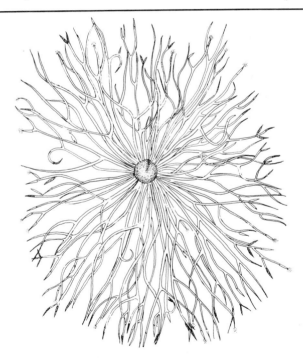

Fig. 18. Growth of the water mold *Achlya* on a hemp seed (Redrawn from J. R. Raper).

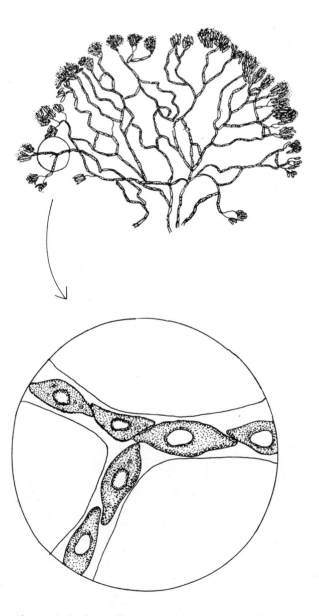

Fig. 19. (*Above*) A feeding edge of a colony of *Labyrinthula*; (*below*) a high-power view of a portion of the net, showing the individual cells.

A number of multicellular colonial protozoa that attach to the substratum have evolved independently. One might presume, in the case of the colonial ciliates, that they stem from a sessile, *Vorticella*-like ancestor, and have become multicellular by failure to separate after division (Fig. 20). What advantage this provides them is very hard to see. It is true that some, such as *Zoothammion*, have become highly complex and internally coordinated, but this must have been a later development. One is tempted to assume in this case that the accident, the mutation, that resulted in an inability of the daughter cells to separate may have been selectively neutral and was retained merely for lack of

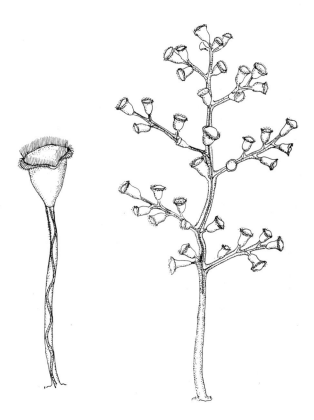

Fig. 20. *Vorticella (left)* and *Zoothamnium (right)*.

pressure to get rid of it. The coordination could have appeared secondarily; some colonial ciliates lack it totally. This genetic-drift argument must remain provisional as long as we cannot demonstrate specific advantages for a particular environment between the single-celled *Vorticella* and the multicellular *Zootham-mion*. To say that the selection pressure is neutral is an impossible argument to prove, for, like all negative evidence, it stands until it is convincingly demonstrated to be wrong.

The same argument arises among many other protozoa. D'Arcy Thompson[26] was especially eloquent in defending his belief that the vast diversity of the shapes of the shells found among radiolarians and foraminiferans was not due to selection, but was the result of permutations of molecular forces, much the same as the cause of the variations in snowflakes. He could not agree with the view of Rhumbler and others that the shells differed in strength and that there was a selection pressure for this strength which would explain the changes in the form of foraminifera through evolution: "But in days gone by I used to see the beach of a little Connemara bay bestrewn with millions upon millions of foraminiferal shells, simple Lagenae, less simple Nodosarieae, more complex Rotalieae: all drifted by wave and gentle current from their sea cradle to their sandy grave: all lying bleached and dead: one more delicate than another, but all (or vast multitudes of them) perfect and unbroken. And so I am not inclined to believe that niceties of form affect the case very much: nor in general that foraminiferal life involves a struggle for existence wherein breakage is a danger to be averted, and strength an advantage to be ensured." Again the question of whether or not there are different niches in the sea which have very slightly different selection pressures that produce this enormous variety, even among living radiolaria or foraminifera, is a question impossible to answer at the moment. With the new work on small changes in protein structure, we are becoming used to the fact that neutral (or almost neutral) mutations could exist, and some population geneticists[27] lean heavily on them in their arguments.

It must be remembered that foraminifera are primitive multi-cellular organisms, for they are multinucleate during much of their life history. They have an alternation of haploid and diploid generations, and the shell size and shape in these two generations are different. The interesting point about these shelled amoebae is that if they are measured in sequential deposits over long periods of geological time they show marked progressive changes of various sorts. (Fig. 21). This can only mean that some aspects of their shape must have had some adaptive significance, however slight, but it is impossible to know what those advantages might have been.

The greatest variety of colonial forms in primitive multicellular aquatic organisms is found in photosynthetic species. J. R. Baker[28] has made the point that this variety is possible because no communal feeding apparatus is needed; being photosynthetic, they need only to catch the sun. This again raises the question whether or not each form of colonial alga is specifically adapted, or do we again have an example of a form in which the daughter cells failed to separate, but the selective advantage is nil. It is hard to believe in this case that the selective neutrality spread indefinitely, because soon the size and the complexity of the algae became such that they must inevitably have come under strong selection pressure. One need only look at the living volvocales to see the possibilities. *Pandorina* and *Eudorina* are relatively small, while *Volvox* is large and more intricately organized (Fig. 22). The cells of its northern hemisphere differ from those of its southern hemisphere (which bear the reproductive cells): there is a differentiation to the whole multicellular colony. This specialization is reflected in the fact that if a cell of *Volvox* is removed, it fails to regenerate; a cell of *Eudorina* will produce a miniature colony, while cells of the simpler forms are all capable of complete regeneration.[29] Consider finally the fact that the asexual daughter colonies of *Volvox* are kept internally, protected in the mother colony in a primitive kind of womblike existence. (There can even be two generations, one inside the other, a phenomenon first observed by van Leeuwenhoek.) Just as in the form progression in the foraminifera, it is certainly reasonable to assume that in this series of colonial organisms there has been a selection for increased integration. It cannot be proved; we do not even know

101

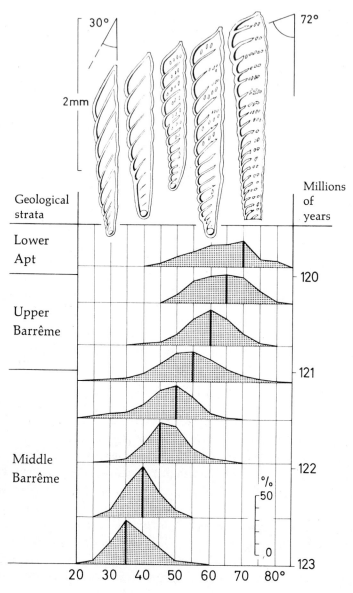

Fig. 21. Evolutionary trends in the foraminifera. The angle of the compartments of *Vaginulina procera* Albers has slowly shifted over a period of time as one examines specimens from successive deposits. [From F. Bettenstaedt, *Mitteilunger der Geologische Staatsinstitut, Hamburg, 31* (1962), Figs. 4 and 5.]

their early history, for, in contrast to the foraminifera, there is no palaeontological record. All the existing smaller forms might be recent in origin. This raises the more important question of why they all exist today. Our answer must be that each in its own way is adapted to some set of aquatic conditions. They all have resistant, dormant stages, and they all can wait until those perfect conditions arise. If a particular condition is the same for two forms in the volvocales series, we must assume that at least under those conditions they can coexist, although that does not necessarily mean that they do so under all conditions.

The difficulty in this discussion of the evolution of the volvo-

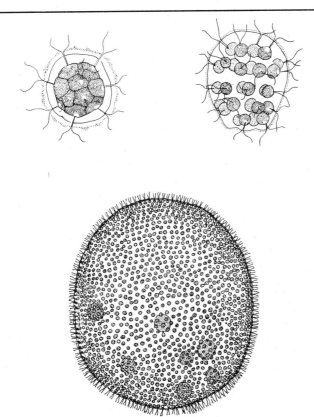

Fig. 22. Three different-sized colonies of Volvocales: (*upper left*) *Pandorina*, (*upper right*) *Eudorina*, (*below*) *Volvox*.

cales and the foraminifera is that we have not been able to pin down the source of selection. We have suggested size and integration, but these are only secondary qualities. The basic ones are those connected with the efficiency of the intake of energy and the use of that energy to make more offspring. In our previous examples this meant success in feeding and dispersal. It is unclear whether feeding efficiency is a significant factor in foraminifera, but dispersion is a possibility. The number of gametes or the number of asexually produced microcolonies will be greater in the larger forams, and it is conceivable that under some circumstances a longer generation time and a larger discharge of reproductive bodies at key moments would be advantageous. In the case of the volvocales, perhaps the fact that a daughter colony is built up to a larger size before liberation might, under certain circumstances, enhance its chances of survival and therefore of producing offspring. The arguments are bound to be hypothetical and therefore unsatisfactory to an appreciable degree.

There is some debate among zoologists as to the origin of metazoa.[30] Many hold the traditional view that they came from some sort of flagellate ancestry, although others believe that the ciliate protozoa gave rise to the acoel flatworms and that this was the beginning. Both schemes (or neither) may be right, and this is not the place to examine the evidence. Instead, let us take a brief look at sponges, which are primitive both in structure and in development.

Sponges are not photosynthetic and therefore require a mechanism for capturing particulate food. Unlike the myxomycetes or *Labyrinthula*, they do not crawl about the surface, breaking down and sopping up the products of detritus, for they are sessile. This very fact means that they must bring food into themselves from the surrounding water, a problem that also faces a sessile ciliate such as *Vorticella*. Both manage it by waving their cilia or flagella and setting up water currents, which brings the food particles (and oxygen) to a place where they can be engulfed by the cytoplasm. Sponges are larger than ciliates and they have the problem that only a fraction of the cells are flagellated feeding cells. This means that the feeding efficiency of the collar cells must be great if they are to maintain the whole sponge. We are indebted to G. P. Bidder[31] and his famous paper for our appreciation of the efficiency of a sponge. He showed that a smaller leucon type (slightly larger than one's finger) can, because it is hydraulically efficient to an amazing degree, send out a jet of its waste water some 40 centimeters. This is done by building up the pressure in many small flagellated chambers, and by having the apical osculum just the right diameter (much as one controls the nozzle size on a garden hose) so that the water shoots out the maximum distance possible for that particular pressure. (Bidder was also able to calculate the muzzle velocity at the osculum. Holding a pipette with a rubber bulb containing carmine to the side of the sponge, he noted that the dye came out in distinct waves (Fig. 23). After some puzzlement as to the possibility that the flagellate cells might be working in a rhythm, he suddenly realized that the effect was caused by his own pulse as he held the bulb of the pipette between his thumb and finger. By measuring the time interval between heart beats and the distance between the waves of carmine at the oscu-

Fig. 23. G. P. Bidder's method of determining the jet velocity of a sponge.

lum, he could calculate the speed at which the water left the sponge.)

Sponges with such an efficient pump are ones that live in still water; the pumping prevents the reuse of the same water. Some forms increase the efficiency of preventing re-use by being stalked. Young sponges, which have not yet grown enough collar cells and flagellated chambers to build up the necessary pressure, will have a long tube or osculum chimney that leads the used water off some distance from the small sponge. It is possible to see, in the different types of sponges that exist today, what must be the evolutionary sequence leading up to this efficiency. In ascon sponges there is one simple chamber; in sycon sponges the surface area is increased by the formation of radial canals; and in the leucon sponges, there is a massive network of flagellated chambers (Fig. 24). This is a trend toward a relative increase in the area of the flagellated surface, which provides the power for the pressure, and control of the size of the flagellated chambers, the collecting tubules, and the excurrent osculum, so that the

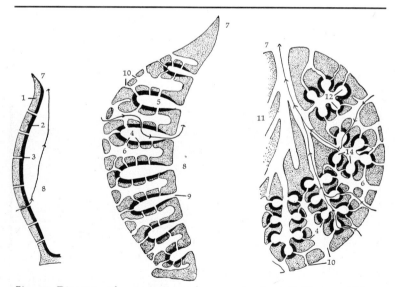

Fig. 24. Diagrams of various types of sponge structure: *(left)* asconoid type; *(middle)* syconoid type, with cortex; *(right)* leuconoid type. Choanocyte layer in solid black, mesenchyme stippled. 1, mesenchyme; 2, choanocyte layer; 3, incurrent pore; 4, prosopyles; 5, radial canal; 6, incurrent canal; 7, osculum; 8, spongocoel; 9, internal ostium; 10, dermal pore; 11, excurrent channel; 12, flagellated chamber; 14, apopyle. [From L. Hyman, *The Invertebrates, Protozoa through Ctenophora* (McGraw-Hill, New York, 1940), Fig. 79; used with permission of the McGraw-Hill Book Company.]

pressure and hence the distance the used water is propelled are optimal. The perfection of the feeding system in sponges is an ideal demonstration of how size and efficiency in feeding by increase in complexity go hand in hand; natural selection has proved unsurpassed as a hydraulic engineer, and all because more efficient feeders are more efficient reproducers.

If we look at reproduction in sponges in terms of dispersal, it is not clear how any one sponge might have an advantage over another. In sexual reproduction the larvae swim away to start a new individual, so in this case dispersal is by currents and the swimming capabilities of the amphiblastula larvae. In some marine sponges and in most freshwater sponges there is also an asexual reproduction by gemmules. In this case there is a remarkable adaptation to seasonal changes, for some gemmules need to be vernalized, that is, cooled for an appreciable period, in order to germinate. In other species of sponges the gemmules can germinate immediately. There is even a species that produces both types simultaneously: by having both strategies, it can take advantage of either warm or totally frozen winters.[32] So advances in reproduction efficiency in sponges lie in the motility of the larva for effective dispersal and the ability of resistant gemmules of different types to cope with seasonal variations in temperate zones.

In each example of the origin of multicellularity we see two principle ways in which selection has increased the success of reproduction. One of these is to increase the efficiency of feeding; more energy means more reproduction. The other is to improve the method of dispersal, so that the organism finds new food sources and again is able to divert more energy into reproducing itself. Finally, there are many secondary adaptations that also permit this energy-for-reproduction pathway to occur, such as the appearance of resistant stages to cope with the fluctuating environment.

The origin of social organisms

There are so many different kinds of social groupings of organisms that, as a first step, it is necessary to have some sort of classification or categorization. I agree with E. O. Wilson that it is useful to have a very general definition of a society, such as "a group of individuals that belong to the same species and are organized in a cooperative manner."[33] This cooperation involves communication among the individual organisms. The difficulty is that because this definition is so general it automatically includes relatively solitary animals which may come together in pairs for a brief mating season, and it includes all animals that have parent-offspring relations. At the other end of the scale, some of the social insects have extraordinarily elaborate and structured societies that by their very complexity are another class of phenomena. Again following Wilson, the students of social insects have used the term "eusocial" for this highest category. Eusocial insects have three main characteristics: (1) they not only care for their brood, but do so by cooperation among a number of individuals; (2) there is an overlap of generations so that at least two generations are able to share the work load; (3) there are castes, that is, there is a division of labor of the cooperative individuals (with the production of sterile and sometimes of polymorphic workers).[34] There are among insects, other invertebrates, and many vertebrates all sorts of intermediate states, which have one or two of these characteristics, and therefore these animals fall short of being eusocial. There is a variety of terms to describe these intermediate levels, but for the sake of simplicity in this discussion we shall use the general term "social" and the one specific term "eusocial." There is, however, a particular type of primitive social organism in which the individuals do cooperate, but are not separate; they are physically attached to one another. To distinguish this category we shall follow the common usage of calling such organisms "colonial," as in the colonial hydroids. (This should not interfere with the use of the term insect "colony," which is a family of insects that live in one group.)

Because there is cooperation among individuals, we have a grouping that is to varying degrees integrated, and therefore, to varying degrees, new properties of the whole colony or society emerge. As before, there is nothing mystical about these new properties; they are inevitable consequences of the integration. It is especially important to stress this point because there has been a considerable amount of misleading writing on this very score. Clearly, these new qualities must have had selective advantages, for they have appeared independently many times in the course of evolution, and furthermore there have been numerous independent trends toward increased integration, leading ultimately to instances of eusociality in termites, wasps, bees, and ants. The more we learn about social organisms, the greater the evidence is for repeated, separate origins of social groupings.

If selection favors, under some circumstances, social organization of animals, our first question concerns how the information is handled genetically, how it is stored, and how it is varied for a group of closely knit individuals. The interesting point here is that the genetic system of all animals and plants was originally devised for single cells. There was, however, a selection pressure for multicellularity, and the genetic system had to be stretched in its capabilities so that it could remember and manage a multicellular life cycle. Now we are making a further demand on this unicellular system of inheritance; we are asking that it govern the cycle of a collection of separate multicellular individuals that are in a coordinated group. In the case of those organisms where the individuals in the group are physically attached, such as the colonial hydroids, there is a direct continuity of the original egg nucleus to all the individuals. In other words, a hydroid is genetically a clone, and therefore is subject to selection in the normal manner. This does not, of course, explain how mechanically one can get a division of labor between the individuals in the hydroid colony, a matter that we shall consider in Part II of this book. But it does say that the egg cell contains the information on how to manufacture not only an individual, but also a group of individuals that bud off and that can themselves show differences in structure (for example, feeding polyps and reproductive polyps). The genetic problem here, from both the point of view of storage and the point of view of gene action, is basically no different from what is found in separate multicellular forms. However, the fact that increased complexity has been achieved in this particular way is of unusual interest and also valuable in considering the mode of gene action.

A different problem exists among social insects. The first person to appreciate this fully was Darwin, for at one time he felt that the presence of large groups of sterile workers might present an insuperable difficulty for his theory of natural selection.[35] Ultimately he came to realize that the core of the problem was genetic and that the essential inheritance was possessed by the fertilized queen. The insect society is often one large family in which many of the offspring are sterile. But the storage of information remains in the germ cells of the queen; the life cycle is generated by the fertilized egg nucleus that gives rise to the queen. She, in turn, has a copious supply of offspring which are sterile workers that help her in food gathering, brood care, and nest protection; but ultimately the cycle is complete and she generates new fertile reproductives, new males and queens to begin new colonies.

From the point of view of selection, the key issue is that the sterile workers are entirely altruistic; they do not perpetuate themselves. The male and the queen alone pass on the genetic information, and the large production of workers must somehow increase the reproductive success of the king and queen, despite their own dead-end sterility. As before, by reproductive success we mean contribution to subsequent generations. The cost of manufacturing all those nonreproductive individuals must be more than balanced by advantages in reproductive success. We are, along with Darwin, continuing to consider only those cases of the most advanced eusocial insects in which a sterile caste exists; if we can understand the extreme case, those with less advanced social organization will automatically be included.

The concept of the cost of reproduction is again important here. In a tree, all the leaf, stem, and root tissue dies ultimately, while a small fraction of the total tissue forms flowers that produce gametes which lead on to the next generation. In the case of colonial hydroids, the parallel is especially close, because, just like a leaf, a feeding polyp is from the same clone and is physically attached to the reproductive polyp. In both cases selection has favored a division of labor (related to the size increase) which results in a high cost of reproduction by producing auxiliary, non-reproductive tissues. In the eusocial insects the difference is that the cost of reproduction is largely paid for by a large mass of sterile offspring.

The important advance in this subject of genetic theory of insect societies comes from the work of Hamilton.[36] He has also begun with the point that in order for an altruistic trait to appear in evolution, the sacrifice in fitness of an individual (the cost) must be compensated for by an increase in the fitness of the group. But the latter is significantly affected by how closely related the individuals may be genetically. The peculiarity in the hymenoptera is that the males are haploid and the females, including the sterile workers, are diploid. This means that all the workers are very closely related to one another: they receive the identical set of genes from their common haploid father, and they share one half of their genes from their mother. Therefore a worker shares three-fourths of her genes with any other female worker, while a queen or a fertile worker would share only one-half of her genes with her offspring. Because of the haplodiploidy, workers in hymenopteran eusocieties are more closely related to one another than to their parents. It would be advantageous, therefore, for a worker to help and care for its siblings, which would include the new sisters produced by the queen; in fact it would be more advantageous for the fitness of the family than if the queen looked after her offspring, or if any worker became fertile and did the same for her own issue. The advantage to family fitness comes entirely from this difference of degree of relatedness.

It is difficult to grasp the rationale behind Hamilton's theory, and a statement from Wilson may be helpful: "The following example should make the relation intuitively clearer: if an individual sacrifices its life or is sterilized by some inherited trait, in order for that trait to be fixed in evolution it must cause the reproductive rate of sisters to be more than doubled, or that of half sisters to be more than quadrupled, and so on."[37] This

111

example would apply to ordinary diploid organisms, but in the case of sisters of haplodiploid hymenopterans, the rate of reproduction would only have to be increased more than four-thirds; this is the advantage in producing a sterile worker class found in haplodiploid organisms as compared with ordinary diploid organisms.

The cost of reproduction is decreased if the sterile, dead-end units are more closely related genetically. We shall call this the genetic reproductive cost. The extreme at one end of the spectrum is a multicellular organism or a clone. Here all the units, either cells or attached organisms in a colony, are genetically identical. Therefore there is *no* genetic reproductive cost in this case, but merely a cost in the expenditure of the energy involved in making all the nonreproductive tissue. This is, however, a real cost, and, as we have seen, the production of nonreproductive structures must have a positive selective value if they are to persist. At the other end of the scale there are social groupings of diploid families of which the most advanced examples are found among the termites; in fact, they are the sole example of a diploid eusociety. Eusociality apparently arose only once in the termites, while it arose independently numerous times in the haplodiploid hymenoptera, lending futher support to Hamilton's theory.

We have so far discussed social groups composed of clones (in which there is a complete sharing of genes by attached individuals) and family societies (in which the separate individuals share either three-fourths or one-half the genes with their immediate relatives). Now we shall turn to societies made up of a genetically diverse group or population. None of these are eusocial, but among insects and other arthropods there are remarkably complex societies made up of a collection of diverse genotypes. A good example, also described by Wilson, is the spiders that have communal nests.[38] One species, for instance, found in Venezuela, has about a thousand individuals that cooperate to build one large nest, and they hunt in a pack, dragging the prey into a suitable area within the net. There is even a social spider found in Mexico in which a number of females appear to lay their eggs in a communal egg sac.

All vertebrate societies fall in this category of a group of genetically diverse individuals. There may, of course, be some relation, for there will still be families within the group, but there will also be increasingly distant relatives, depending upon the natural history and behavior of the society. For instance, in monkey clans (such as howler monkeys or baboons) there are groups that move as a unit and only rarely does a male from one group enter another; therefore there must be a considerable amount of inbreeding. The same situation is true in wolf packs, while the less well-organized colonies of nesting sea birds or reproducing fur seals will be larger and less tightly knit, and therefore will have greater genetic diversity.

Again in all these instances there is evidence of altruism, although not in the form of sterile workers. The sounding of alarms or the posting of sentries are examples of behavioral altruism. In most of these cases it can be effectively argued that initially the character was attained in perfecting the care and protection of offspring, but in subsequent evolutionary change it became a form of altruism that benefited the whole group, not just the immediate relatives. In this instance the sacrifice made is not mandatory sterility or certain death, but something intermediate, for the animal giving the alarm is exposing itself to danger and increasing the chance that it may be sought out and destroyed. It is merely a question of the probability of death in performing the altruistic task. Here again, whatever the cost of this sacrifice may be, it must be more than compensated for by the selective advantage to the animal's own genome. The more closely related the individuals in the group are, the more the genetic cost of reproduction is decreased. If the trait is preserved in the genes of the group or the population, each individual will benefit by better protection and have a greater chance of producing offspring. Therefore those individuals with alarm-giving genes will be favored, since their own offspring have a better chance of survival. If they include close relatives as well in the protection system, this will more quickly perpetuate the desirable genomes and the alarm mechanism will spread to larger groups. This again emphasizes that it is genes that are the object of selection, and they spread through the population. The easiest method of such spreading is

113

close inbreeding or direct kinship. But no matter whether the gene is prevalent or rare, if it is altruistic and requires danger and sacrifice on the part of individuals that possess it, the danger must be compensated for by the advantages it provides in favoring reproductive success. If the danger means certain reproductive death (as in sterile workers), the need to compensate is automatically greater.

Before we leave the genetics of social organisms, there is one important point to be mentioned. Phenotypic variation, variation that is not inherited, plays a significant role in animal societies The most obvious case is polymorphism among social-insect castes. It is well known that in most instances the differences between workers, even the differences between queens and nonreproductives, are entirely environmental. The genetic endowment of each is the same, and this endowment permits the individual to respond to external conditions in a number of ways. In some, if not all, vertebrate societies the peck order of dominance is probably due to such phenotypic variation. It may even be an important factor in giving the social group a stable structure. There are examples of what we previously called range variation in which what is inherited is the ability to vary within a defined range. The fate of any one individual (or the frequency distribution of the variant individuals within the range) can be influenced by environmental factors (such as food or pheromones). If there is no such external influence, one would expect the frequency distribution of variants to form a normal distribution curve within the inherited range.

We have discussed the origin of animal societies in terms of the cost of altruism; now we shall look into the question of what factors might have been responsible for the societies' arising in the first place. We have constantly stressed the need to compensate for the cost, and here we are asking the more important question: what are the benefits, the selective advantages, that might have been responsible for the formation of these societies? The first way one can look at this problem is to examine the evidence of how they arose initially. The most abundant and varied source of information comes from social insects, and we shall again rely on E. O. Wilson's excellent discussion.[39]

There are two principal avenues of becoming social, both of which are found in many groups of primitive or presocial insects and social spiders. One of these avenues is the care of the young; the other is communal feeding. The examples of insects that care for their young to varying degrees are numerous. There appear to be two main conditions that especially favor such offspring care. A number of primitive social forms, such as dung or carrion beetles, have adopted what Wilson calls a "bonanza" strategy. When they are fortunate enough to find some dung or decaying carrion, they immediately lay their eggs, and presumably there should be no need for further care. However, it is important to keep other dung and carrion eaters away, so the adults perform this defensive function, and if they manage it successfully, they are assured of raising a very large brood. Care of the young is also favored in those species that live in a very uncertain and hazardous location. There is a staphylinid beetle, for example, that inhabits the intertidal mud. If the mother does not constantly keep at work in the tunnels, the brood is soon killed by lack of oxygen. In this case the enemy is the physical environment, the tide, rather than a competing species, but in both cases the selective advantage of offspring care is the protection it provides.

The second avenue of achieving sociality is through communal feeding. A good example cited by Wilson is the sawfly that lives on the jack pine. When the larvae are very small, having just hatched from the egg, they cannot individually chew through the tough cuticle of the pine needles. By forming a tight group and concentrating their combined efforts in one spot they succeed and gain nutriment. Another good example is the social spiders mentioned previously which have a large communal net and by mass attacks can presumably capture large prey. It should be added that many aggregations of animals are involved with mating: swarms in many insects, hauling out on the beaches of the Pribilof Islands in fur seals, and so forth. These groupings may be quite ephemeral, and they are in general less important as a socializing influence. Communal feeding is a far more significant factor.

The point was made that termites became eusocial only once

115

in evolution, originating from solitary cockroaches, while the state was achieved a number of times independently in the hymenoptera, the reason perhaps being that the latter are haplodiploid while the termites are diploid. Whether this is the correct interpretation or not, the fact that it was a rare event even in cockroach history may indicate that it was a difficult state to achieve. L. R. Cleveland, who did so much work on the cellulose-destroying obligate symbionts of termites and the wood-eating cockroaches (which are believed to be their ancestors), made the suggestion that the rise of this symbiotic association was the essential preadaptation for the rise of the social state in the termites.[40] The termite cannot digest wood without its intestinal flagellates. In order to retain its wood-destroying ability, it must retain the family infection of flagellates. At each molt the insect loses its complement of flagellates and must become reinfected by licking other termites. Because it must have other termites to lick, it must be in a social group. This kind of feeding cooperation would have been a powerful selective pressure for tightly knit social groupings, and indeed such cooperation is found in the integrated families of the ancestral wood-eating roaches. (The interest in the story does not end here, for the whole life cycle of the flagellate, including its modified meiosis, is geared to its symbiotic role, even to the point where the ecdysone, or molting hormone, of the termite host stimulates the meiosis of the intestinal flagellate. Evolutionary change has affected both partners.)

If we look for examples of primitive social organization among vertebrates we find the same two factors: care of the young and communal feeding. Both play a significant role and either one could have led independently to social groupings. One of the important features of vertebrates in general is the increase in the prevalence and the degree of offspring care. That this should give rise to complex bonds between offspring and parent is not surprising: they respond to each other in such a way that the young can be fed and protected through parental guidance. This is no place to review the extensive studies on the kinds of communication between adults and their young; all I want to do is to remind the reader that they exist. There are probably a number of separate origins of such behavior extending beyond an isolated family; let me cite two examples. J. H. Cook[41] points out that a nesting colony in which many birds have their nests close together affords considerable protection for the young birds. Either by the colony's being located on an inaccessible cliff or an offshore island, or by sheer number of individuals, predators will be kept away. There is also the well-known effect described by Fraser Darling,[42] who showed that in a large colony of sea birds egg laying was closely synchronized by mutual stimulation, with the result that there was a shorter total period of vulnerability to predators because of a momentary surfeit of prey, while a small, less synchronized colony was open to predation for an extended period. Protection of the young can therefore be supplied not only by the attention of the individual parent, but by having special collections of offspring-raising parents, providing a rudimentary sociality that extends beyond the family.

The advantage of communal feeding can be seen in a number of carnivores that hunt in packs. A. Murie's[43] description of the flanking techniques of hunting packs of wolves in Alaska provides a perfect example. The wolves literally drive the prey into a trap and they all share the prize.

As either arthropod or vertebrate societies become more complex, there are some clear evolutionary trends, each of which must be adaptive. It is easiest to discuss these in the frame of the three conditions of eusociality: (1) cooperative care of offspring, (2) overlap of generations to perform tasks, and (3) formation of castes in a division of labor.

(1) There is no need to dwell on the advanced cases of cooperation among sibling workers in termite and hymenopteran eusocieties. There is almost a continuous spectrum between the extreme cases and examples where a queen shares the workload with a few unspecialized workers. Cooperative behavior of a very general sort can be seen in different vertebrate societies. Many monkey groups or clans have an alarm system for danger and this helps all the young. One of the best examples is the formation of crêches among penguins. When the young are large enough to be able to wander about the windy slopes without freezing

117

to death, the fasting parents will go to sea to feed and return with food for their young. These young are herded together in one large group by a mere handful of parents while their neighbors feed. In all these cases, both insect and vertebrate, the selective advantage is to feed and protect the offspring in the most efficient manner, so that for the energy expended, the returns in producing offspring are high.

(2) In tracing the evolution of the overlap of generations, where two generations perform communal tasks, it is helpful to look at the founding of a new colony of a social insect, such as an ant. The queen may wander off alone and in the beginning she lays the eggs and takes care of the young all on her own. When the first minute workers appear, they will go off and forage and help their mother in rearing the next batch of offspring. From that moment on there will be two overlapping generations of workers, one helping the other. In the more advanced species the queen will soon stop caring for the offspring and leave the matter entirely in the hands of her older daughters; she concentrates her attention on the serious business of making and laying eggs. Among vertebrates there is some overlap of generations in a number of societies. Wolfpacks may be made up of two or more generations, as are the troops of numerous social monkeys. These cases are not so rigid in their distribution of tasks as eusocial insects are, but there is a clear overlap of parent and offspring performing certain tasks together.

(3) The third character of eusocieties, namely, castes, is found only in the social insects. Looking at the origin of castes, Wilson makes the interesting point that in a number of primitive hymenopteran social groupings the division of labor is often achieved by a dominance sequence between the queen and the workers (which she may closely resemble). This is the case, for instance, in the common brown paper wasp, in which only the queens survive the winter. Usually a new colony is started by a single queen. "However, by the time the first brood appears in late June, the majority of the foundresses have been joined by two to six auxiliaries—overwintered queens who for some reason have not managed to start a nest of their own. These individuals

are usually subordinate in status and reproductive capacity to the foundresses. Their subordinance is expressed behaviorally in overt ways: The auxiliaries assumes submissive postures, undertake food-gathering flights and regurgitate to the dominant foundress, and defer to the foundress in egg laying. The foundress not only prevents her associates from laying eggs, she also eats their eggs when they occasionally sneak them into unoccupied cells. In time the ovaries of the subordinates regress."[44] As Wilson points out, these primitive associations are generally managed with a considerable amount of aggression, while in more advanced societies the various castes all live peacefully in harmony, each pursuing its own tasks without interference or intimidation. The idea is that the caste system evolved first by the establishment of roles through overt aggression, and later this was replaced by a more codified division of labor.[45]

It is interesting that in vertebrate societies the same spectrum is found, from aggressive interrelation of individuals to placid ones. In monkeys, baboons show a sharp dominance hierarchy based on what appear to be sustained threats or open aggression, while the howling monkey has an equally well-organized clan but the individuals are gentle with one another and there is little overt evidence of a tyrannical peck order. It is less clear in these monkey societies that one type is more primitive and ancestral than the other; it is equally likely that they arose independently in response to different factors in the environment.[46]

No matter how they arose, and no matter how they are constituted in societies today, so far as we know, many of these instances of division of labor are, as we said previously, somatically or environmentally determined. The caste or the rank of dominance of an individual of either sex is not inherited *per se*, but only the ability to vary and respond to certain environmental conditions such as the amount and kind of food provided during growth.[47] However, here we are not concerned with the mechanism of this division of labor, but simply with why and how it arose. In this we are again indebted to Wilson, who studied caste formation in ants and analyzed the relative proportions of the castes in terms of optimization theory. What particular division of labor will most efficiently and effectively permit the ant family to reproduce successfully?

The first step in this analysis was taken in 1953, when Wilson[48] showed that polymorphism could be described in terms of the relative growth, or allometry, relation of J. S. Huxley. Measuring almost any two physical dimensions (x and y) of the sterile worker ant, he found the following relation between them:

$$y = bx^k,$$

where b and k are constants; k is the ratio of the growth rates of x and y, and b reflects the initial size difference between x and y. Taking logarithms, we have

$$\log y = \log b + k \log x,$$

which on a log-log plot gives a straight line slope of k. Given this relationship, if k is either more or less than 1, the ratio of the two measured dimensions will vary with the size of the ant. This is one way in which polymorphism may be achieved in the workers of a colony.

There are two additional ways. The second is by an alteration of the k with increasing size, producing what Wilson calls a diphasic curve (Fig. 25). In this case the frequency distribution of the size changes will be normal. The third way is by producing a bimodal size frequency distribution in which major (soldier) and minor workers will greatly outnumber intermediate forms. This latter condition is thought to be related to the fact that the allometry curve is triphasic (Fig. 26).

One of the most striking aspects of social insects with distinct classes of sterile workers is that the numbers of the different types keep a constant proportion. When we consider mechanisms in the second part of this book we can examine how this is achieved, but here we are solely concerned with "why." The matter has been approached by Wilson[49] using optimization theory in the following way. Very briefly, he takes into account the temporal division of labor, the sequence of tasks any one worker will perform at different times during her lifetime, and assumes an overall contribution (task) for a particular worker class. The colony will, for its competitive well-being, need protection and food gathering; these colony needs he calls contin-

120

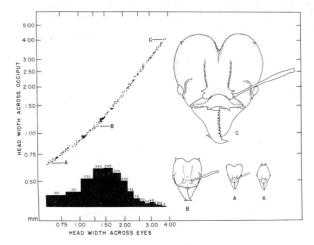

Fig. 25. Workers of the fungus-growing ant *Atta texana* show a polymor-phism seen in head sizes *A, B,* and *C.* Their allometry is diphasic, meaning that the relative-growth ·curve changes slope (at point *B*). This change is an adjustment to accommodate the great size variation shown by the species. If the slope of the lower segment of the curve were as high as that of the upper segment, the smallest workers would have the monstrous head shape (*X*) in the lower right-hand corner of the figure. [From E. O. Wilson, *Quarterly Review of Biology*, *28*, 136–156 (1953).]

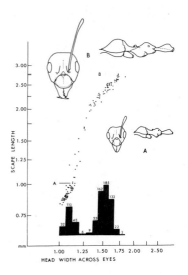

Fig. 26. In *Oecophylla smaragdina* the allometry is triphasic, that is, three different slopes are shown by such allometric characters as scape length taken as a function of body size (in this case represented by head width). The size-frequency curve is bimodal, with majors predominating. The heads, mesosomas, and petioles of selected minor and major workers are also shown. [From E. O. Wilson, *Quarterly Review of Biology*, *28*, 136–156 (1953).]

gencies that must be dealt with by the workers. The question is what size of the worker population and what mix of different castes will provide the optimum reproductive success of the colony queen. Wilson represents graphically a simple case in which there are two tasks and two castes. The ideal or optimal mix is where the curves of the two tasks intersect (Fig. 27).

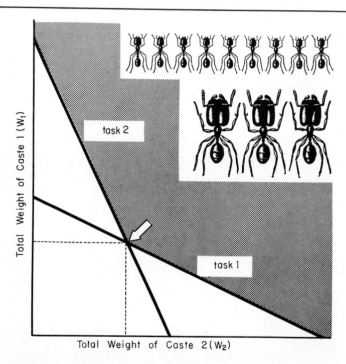

Fig. 27. The simplest possible general form of the solution to the optimal-mix problem in evolution. Two kinds of contingencies ("tasks") are dealt with by two castes. The optimal mix for the colony, measured in terms of the respective total weights of all the individuals in each caste, is given by the intersection of the two curves. Contingency curve 1, labeled "task 1," gives the combination of weights (W_1 and W_2) of the two castes required to hold losses in queen production to the threshold level due to contingencies of type 1; contingency curve 2, labeled "task 2," gives the combination with reference to contingencies of type 2. The intersection of the two contingency curves determines the minimum value of $W_1 + W_2$ that can hold the losses due to both kinds of contingencies to the threshold level. The basic model can now be modified to make predictions about the effects on the evolution of caste ratios of various kinds of environmental changes. [From E. O. Wilson, *American Naturalist, 102,* 41–66 (1968).]

By varying the number of tasks, the environmental conditions, the degree of specialization of the castes, and a change in efficiency of one caste, it is possible to make various predictions as to how the optimal mix might shift. From these theoretical considerations it can be shown that under certain conditions there will be an evolution toward an increase in the number and specialization of the castes. (Other conditions could favor the reverse.) As this specialization or task efficiency rises, the need for large numbers of individuals of that caste will decrease, so that both the proportions and the size of the colony are constantly being affected by selection.

The final point I wish to make about social organisms is that some of the more general advantages of size increase are again factors that must play a fundamental role in the adaptive advantages of social existence. Examples of this have already been given, for instance in communal feeding, but there are others. The most striking one centers around our previous argument that progressive internalization, or isolation from the environment, has been an important adaptive strategy in higher multicellular animals, and homeothermy was given as an example. A close parallel may be found in social insects; thus, individual honeybees are poikilothermic, but the hive is homeothermic.[50] By fanning, the bees can cool the hive by evaporation of water, and by muscle contraction, they can raise the temperature. In this way they keep their hive between 34.5° and 35.5°C all through the foraging season, from late spring to the fall, even on very hot, sunny days (up to 70°C) and on extremely cold nights.

Summary and conclusions

For each major level, the cell is the minimum unit of the life cycle. In the evolution from prokaryotes to eukaryotes, the basic cell increased in size and complexity, an innovation that permitted a far greater range of form in unicellular and especially in multicellular and social organisms. The molecular components of the minimum cell have many inner molecular replacement cycles, yet the end result, the cell, remains in a steady state in the sense that it never goes below this minimum collection of information (both DNA and DNA derived). The information is for the elaboration of the phenotype in a repetitive life cycle that can become, through natural selection, remarkably diverse if one looks at all the unicellular, multicellular, and social organisms.

There are two main themes in the first half of this book. Each, in its own way, says something about the larger significance of the development of living organisms. We have asked such questions as why development occurs, how it arose during the course of evolution, and what is its relation to all other aspects of life. We have tried to put the process of development in its proper setting, its proper perspective.

One of these themes, which has been developed in this chapter, is that at each level of complexity we considered what are the factors that favor an increase in the cost of reproduction, for we know that the advantages of successful reproduction must outweigh the cost if more elaborate phenotypes are to be selected. The possible strategies for successful reproduction are numerous: it is most convenient to consider them in terms of the gathering of energy resources. Species are either competing for them or they are avoiding competition by using different resources. Success in competition (or competition avoidance) is achieved in three main (nonexclusive) ways: by feeding, protection, and dispersal mechanisms. (1) If two species are eating the same food, in general one will outcompete the other; if they are eating different foods (and exist in a patchy environment), they can coexist and avoid competition. (2) The more effectively a species can protect itself and its offspring, the more individuals will be able to reap the available food resources, an important strategy in competition. (3) Colonizing species depend heavily on effective methods for dispersal; in this way they can outcompete (or avoid) competitors by reaching new sources of food more rapidly and more consistently.

With natural selection, organisms have evolved a wide variety of different mechanical means of improving their methods of feeding, dispersal, and protection. Most of the adaptations are specialized, although it is possible to make at least one useful generalization. Size has changed during the course of evolution, and in particular there has been a steady increase in the upper size limit. For instance, for purely mechanical reasons connected with the principle of similitude, an increase in size will permit an increase in the rate of locomotion, which in turn will affect the ability of the organism to catch food, to disperse, and to protect itself by escaping from predators. For similar reasons an increase in size will make it possible for the organism to insulate itself from the environment and thereby be less disturbed by sudden fluctuations in temperature or other climatic variables.

The other main theme of the first half of this book is another way of looking at the same point. The evolution of an increase in reproductive cost is the evolution of the life cycle (which is development). We have assumed that at first the life cycle was simply a nucleic acid duplication and degradation, and because such a nucleic acid cycle could be influenced by natural selection, there arose protein synthesis, metabolism, and the cell as a structure, a container. This progression has produced two distinct kinds of cycle. One involves many inner cycles that are either asynchronous or so staggered that they maintain the cell as a structural unit in a steady state. It is only by providing this constant structural entity that the DNA can code for all the activities and structures characteristic of living organisms: the DNA products are accumulated over a period of time, so that at any one moment the cell has a complete battery of substances needed for all its functions. The other cycle is the life cycle of which the cell is the point of minimum size. All increases in the cost of reproduction have involved increases in the size, duration, and complexity of the life cycle. It is the life cycle which is the part of the organism that is the prime object of natural selection, and this accounts for the vast array of variations in the life cycles (or developments) that exist among all animals and plants.

Thus far I have attempted the beginning of a macroexplanation or macrodescription of development. Now we must seek an understanding of development in terms of its component parts, that is, a microanalysis.

II

The Molecular View of Development

6

The Synthesis of Substances

The literature on developmental biology that is pertinent and important is extraordinarily large. The main reason for this is that much of what has been written in this century and the latter part of the last century is as modern today as it was then. One need only read Wilhelm Roux, Hans Driesch, T. H. Morgan, and E. B. Wilson to be reminded that this is so. It is true that recently we have had a sudden increase in knowledge of the molecular aspects of development, but that has not put much of the earlier work out of date; it has merely added to it.

This in itself is not such a problem, but the facts and the interpretations of those facts are in considerable disorder. It is as though a library had continued to accumulate pamphlets and books, all (or at least many) of them valuable, for almost a hundred years, but every so often a new librarian appeared with a new system of cataloguing and classifying. As a result the library now has stacks of forgotten papers, vast quantities of scrambled bookcases, and filing cases jammed and overflowing. The disorder and dust are so serious that the latest librarian (the molecular developmentalist) has set up shop in the hallway; at least his files look neat and orderly and new. But with his splendid and exciting new system he has found it impossible to encompass the morass that lies behind him.

It would be misleading to say that I intend to straighten out all those vast tomes in the back room. That would require the patience of an encyclopaedist. Instead, what I would like to do is propose yet another system of classification that, at least potentially, could serve as a reorganization of the entire library. It is definitely not my intention merely to set up another desk further down the hallway. Instead I want to have the system of classification so simple, so straightforward, and so logical that everything (or almost everything) will fall into place.

But before plunging into the particular plan to be proposed here, let us look in a very general way at what some of the "librarians" of the past have done. There are those, such as Hans Driesch, whose perception of the problems was so keen that they felt there were no possible mechanical solutions and so turned to vitalism, the very point where we leave them. There are many who isolated one aspect and by concentration of effort succeeded in illuminating that particular feature of development. Hans Spemann discovered and laid down the foundations of our understanding of induction; Frank Lillie did the same for the biochemistry of fertilization; E. G. Conklin and E. B. Wilson did the same for determinate cleavage; C. M. Child was the first to show the ubiquitous presence of axial gradients. There are many other examples, and in each case one corner of the library was made orderly, but the rest was not touched.

There are those who strove to look at all the issues and there were (and still are) many able developmentalists in this category. Let me very briefly give a rather general classification scheme that fits the views of many and includes our standard modern concept.

In order to go from egg to adult, *growth* is necessary. This is the synthesis of new protoplasm and it manifests itself in the form of cell division and cell enlargement, usually occurring simultaneously, or, I should say, successively. In organisms that increase in size by growth, there is often a shaping of the embryo, and this is generally called *morphogenesis*. (Some prefer to specify those shape changes that are due to the formative movements of cells, using the term *morphogenetic movements*, since

130

one is isolating that aspect of morphogenesis in which there is a cell or protoplasm movement.) In these growing and shaping masses of protoplasm there is ultimately a difference in different parts, a difference in organelle, cell, or tissue structure which it is standard to call *differentiation*. There are variations to these categories, but in this or some allied form they represent the conventional method of organizing developmental processes.

One feature of development is that generally two or all three of these processes are occurring simultaneously. In the formation of the neural tube in vertebrate embryos there are cell growth, cell movement involved in morphogenesis, and the beginning of the differentiation of the nerve cells. A long time ago Needham[1] stressed that it was possible to dissociate these processes one from another by experiments (for example, to have differentiation without growth or morphogenesis) and a vast amount of work has been done since then making use of this dissociability to analyze the mechanism of any one of the three processes. In fact, some organisms, such as slime molds, in their life cycle normally separate growth from the morphogenetic and differentiation stages that follow. But it has always been appreciated that although these are separate concepts, and very useful ones too, they are to some extent rather arbitrary and artificial bins in which to place certain observed phenomena. This is especially true when one wishes to consider any stage of development in which two or all three occur together, a quite general condition. In this instance there are two separate questions: what is the effect of processes on each other, and how does the organism develop? One hopes that the analysis of the first question will help with that of the second, but often the advantage gained is very slight.

If one were to organize all of developmental biology in this way there would be three large areas. But there are problems. One is that these categories serve ideally for multicellular organisms, but they are not so useful in the cycle of a single-celled organism, as J. M. Mitchison has pointed out to me. Growth may still be satisfactory, for there is a synthesis of new substances, but differentiation presents serious difficulties. Is the duplication

131

of some part of a cell, for instance a mitochondrion or a basal body, to be considered differentiation? One certainly considers these cell organelles as differentiated cell structures; the question is, are they permanently so, or is there new differentiation at each division? The word problem does not worry me as much as the fact that there are some appreciable differences between cell and multicellular life cycles and it is important not to obscure these differences by thoughtlessly applying the words devised for one level to all other levels.

A second problem is especially important, for in our three areas we have neglected the most interesting aspect of all, an aspect that transcends all the areas. We do not merely want to know that an organism grows, has formative movements, and differentiates; we want to know how these processes are controlled. An animal or a plant does more than just grow; it grows in a precise way so that in each generation the same shape is produced. The direction and the amount of growth are regulated in some exact fashion. The same may be said of morphogenetic movements; the movements of gastrulation are highly controlled and consistent. It is perhaps most obvious in the case of differentiation. A mammal will not only be properly proportioned in terms of the amount of tissue in various regions of the body (as a result of the control of growth and morphogenetic movement), but also the proportion of the number of nerve cells, muscle cells, liver cells, and so forth will be consistent to a remarkable degree.

It is, of course, obvious and axiomatic that the ultimate source of such control resides in the genome. This is the reason for the great enthusiasm for a molecular approach to development, for we know so much about the molecular basis of genetics. What has happened so far (as we shall see in detail presently) is that there have been great advances in how one gets from a particular gene to the protein it is coded to design, but now we are asking how this protein, or how many such proteins, produce a consistent gastrulation, or a perfectly proportioned leg, or an eye. We even need to have it direct the whole structure and composition of an ant society. The gap between those first gene products and such complicated end results is the gap that lies yawning before us; this is the sensitive spot.

It has been put in such a way as to seem a maximally difficult problem. Let us look at it in a more friendly light. It is obvious that there are many steps, many events between the initial proteins (and all the other structures that come full blown from the previous cycle) and the final events such as we have described. Ideally, what we would like to understand is the mechanics of each step so that we could follow the whole process, like the sequential events that take place in a chemical factory which manufactures some particular substance. Each step must be understood not only in terms of the chemical reactants, but also in terms of where in space and when, in time, the chemicals are in relation to each other. The only difficulty is that in organisms the steps are not obviously laid out, as they are in a factory, and one immediately is impressed by how minute are our living factories, and how fantastically complex; their complexity would put any industrial chemical plant to shame. Nevertheless, some progress has been made and a few links in the chain of events have been identified. The important point is that the events in this series may soon become quite independent of the genome. In our chemical-factory analogy, a particular reaction may automatically occur when two reactants are mixed, but the question of when they are mixed and in what quantities involves control. Valves are opened and closed on a fixed schedule which might be arranged by an elaborate time clock in some central control room. Obviously, if we know how the clock works and how it operates the valves, we have a very much deeper understanding of how the factory works. This analogy is useful for another reason as well. Besides the valve control and the reacting mixtures, there is a third element that is essential to an understanding of the proper functioning of the factory. The positioning of the pipes and the reaction vats, their demensions, in fact their total structure, will affect the operation of the factory. So in organisms we must look for the constituent chemical ingredients, their arrangement in space, and the factors that control their reactivity in space and time.

133

With this brief background we can now plunge ahead and say in precise terms what we are after, what we want to know about development. There is a simple and straightforward way of bringing together the concepts of growth, morphogenetic movements, and differentiation and at the same time of pinpointing the problem of their collective control.

First, let us not forget that reproduction occurs in successive reproductive or copy-making cycles. In these cycles particular substances appear at certain times in specified places. Our aim, which is the aim of all developmental biologists, is therefore to find out (1) what are the substances and how they are produced, (2) when they were produced, and (3) where they are placed. In particular, we wish to know how these dependable and consistent aspects of the copy-making process are controlled and fixed in each generation.

In this very general statement we have included the three processes, and they apply for all levels of complexity. A particular substance may appear at a particular place by growth or by morphogenetic movement. The presence of that substance may in some cases be a reflection of what we commonly call differentiation. Presently we shall examine in detail the full implications of the statement; here let me emphasize that what comes out of such changes of substance production and their placement through time is changes in form, and furthermore changes that go through cycles of alterations each generation.

The usefulness of this statement of the problem of development must now be put to test. We shall systematically examine (1) the mechanisms for the production of substances, (2) the timing mechanisms, and (3) the localization mechanisms, and how each is controlled.

The production of substances

The questions of what the substances are and how they are produced are the very foundations of modern molecular biology. The backbone of all considerations is that, with the help of specific enzymes, DNA makes RNA, which is responsible for the construction of specific proteins. Some of these proteins are the

134

enzymes involved in the synthesis of a vast number of different substances: fats, vitamins, purines, pyrimidines, sugars, and innumerable others. There are a few, such as water and simple salts, that are taken in directly from the environment and used without modification. The cell is a factory which processes energy so that it is continuously manufacturing these substances, continuously metabolizing.

Here we are concerned specifically with the question of how the kind and amount of a substance are regulated. We are interested in the control of synthesis. To begin in a very general fashion, let us list the various obvious control points in the classical synthesis sequence given above.

The main ones are: (*a*) transcription, (*b*) translation, and (*c*) enzyme activity. My plan will be to discuss each one of these individually and then discuss various ways in which these control points may be linked or interconnected. However, the reader must be forewarned that the discussion will be brief and superficial to the degree that a molecular biologist will be quite rightly scandalized. My only defense is that these matters are so well treated elsewhere[2] that another critical review or comprehensive text in this very active field of enquiry is not appropriate in the context of this book. The purpose of the brief sketch here is to show, in as few bold strokes as possible, what are the kinds of phenomena that are encompassed in our present understanding of molecular development. This means that many examples will be neglected, and even more important, I shall not continually lay stress on the tentative nature of many of the conclusions. Let me simply say that although many of the basic facts concerning the regulation of protein synthesis are firmly established, especially for *Escherichia coli*, this is very much not the case in eukaryotic development. There is considerable argument by analogy as to what is known for prokaryotes, and there is even more speculation as to how to account for the differences. It is clear that ultimately each hypothesis must be subjected to the most rigid experimental tests, and these must be different in kind and repeatable for each proposition. This is precisely what molecular developmentalists are in the process of doing at the moment, just as *E. coli* molecular geneticists continue to refine the details of their entire interpretation of prokaryotic molecular genetics.

Transcription

Transcription involves reading off the code of the DNA by producing a *messenger* RNA that forms by template on the DNA and is eventually responsible for the design of specific proteins. This means that transcription is a specific control mechanism in determining what substances are manufactured within a living organism. One could argue that an enzyme also has the property of determining what kind of molecule is synthesized, but each enzyme, and hence its synthetic endowments, are all directly derived from the DNA during the transcription process.

But not all the substances coded in the DNA are transcribed at any one time: what decides the particular moment of the transcription for a gene in a cell's life history? As we shall see further on, there are various external agents, such as hormones, but here we are concerned with more immediate controls. We can begin to understand the question if we look at transcription from the point of view of what in the system can be controlled. The easiest way to see the complexity of the matter is to examine a table that J. E. Varner has prepared on what he calls the possible points of control of enzyme synthesis (Table 1). It is quite obvious that transcription can be influenced in an extraordinarily large number of quite different ways, and it quickly makes one realize how much there is to the question before one can boast of understanding how transcription is controlled.

There is a cruder way of looking at transcription: the control could be either of a negative kind, that is, a suppression of transcription that is occasionally lifted, or of a positive kind, which stimulates active transcription in an otherwise passive genome. The former is, of course, the essence of the famous Jacob-Monod operon hypothesis, in which the gene is ordinarily prevented from being transcribed by the presence of a repressor protein that attaches to a specific region of the DNA and prevents transcription. It is only by removal of this repressor that trans-

Table 1.   Possible points of control of enzyme synthesis.
(From J. E. Varner, *Symposia of the Society for Experimental Biology*, 25, 198 (1971.)

The synthesis of mRNA
  Initiation: changing rate of, by changing
    Number of cells capable of responding
    Number of gene copies per cell
  Accessibility of template
    Affinity of RNA polymerase for accessible template
  Rate of transcription
  Termination
  Release
  Activation
The functioning of mRNA
  Formation of complex with ribosomes
  Movement relative to ribosomes
  Repeated translation
  The synthesis and function of tRNA: specifications of kinds of pro-
    tein and rate of synthesis by
    Iso acceptor tRNA species
    Multiple forms of amino acyl tRNA synthetases
    Cytokinin containing tRNA's
    Other modified bases in tRNA
Polyeptide synthesis
  Initiation
  Translocation
  Peptide bond formation
  Termination
  Release
  Folding

---

cription can proceed; it is activation by means of a double negative, that is, derepression.

There has been a long discussion of the role of histones, and
more recently of phosphoproteins, in the repression and specific
control of transcription in eukaroytic cells. It is known that these
proteins are part of the chromosomes and do affect the synthesis
of certain RNA's in specific ways. The difficulty has been in

137

conceiving of how such a generalized muffling of transcriptive activity could account for all the subtle changes in transcription that must occur during development. We may hope soon to have a better understanding of protein control of transcription in eukaryotes. It is obviously a phenomenon of major, and perhaps key importance; many biologists have based their concept of developmental control on the straightforward fact that only some of the genes are being transcribed at any one moment.[3]

There has been considerable excitement generated by the discovery of the role of sigma factors, which are specific protein components of the RNA polymerase and apparently can affect which genes in the genome are transcribed.[4] The evidence comes from work on phages and especially from interesting studies on spore formation in bacteria. It has been shown that a different sigma factor exists for the synthesis of RNA's used in the vegetative stage from the sigma factor that is active during spore formation. In other words, when combined with a particular sigma factor the RNA polymerase can read only a particular set of genes. In its simplest terms, if this were true for all cells, eukaryotic as well as prokaryotic, then each differentiated cell type would have its special sigma factor.

Although this is an ideal and exciting demonstration of positive control of transcription, there are some problems. First of all, is it a general phenomenon? If it is, then immediately one will want to know how one switches from one type of sigma factor to another. Sigma factors are proteins, so their synthesis must, in turn, involve a polymerase to make their own *messenger* RNA. In the case of spore formation in bacteria, it is known that external factors will promote the change from vegetative growth to spore initiation; do these conditions in some way affect sigma-factor synthesis, and if so, in what way? The theoretical possibilities are immense, but the facts are still wanting.

One approach to the problem of what immediately controls transcription is to examine transcription during the cell cycle. This has been reviewed by J. M. Mitchison,[5] who points out that there are some enzymes ("linear" enzymes) that are produced constantly at all times, except that their rate of synthesis doubles shortly after the DNA doubles by replication. This tells us that at every moment during the cell cycle (except for a short pause after replication) the DNA is transcribed, and furthermore it implies that the amount of enzyme produced is proportional to the amount of DNA. The synthesis of other enzymes ("step" and "peak" enzymes) seems to be totally independent of DNA replication, and do not occur continuously, but in bursts that do not necessarily correspond with the DNA cycle. There are two theories as to why this might be. Most cases are thought to be examples of what Mitchison calls "oscillatory repression." The idea (developed by Pardee and his co-workers)[6] is that an enzyme product has a negative feedback and represses the transcription until the product disappears as the enzyme disappears (owing to lack of replacement by synthesis). In this way there could be a well-timed oscillation of transcription whose frequency might be greater or less than (or equal to) that of the cell or DNA replication cycles. There are a few cases in which the periodicity of the enzyme production could be explained by "linear reading." Halvorson and his associates[7] have produced good examples in yeast: there is one transcription at one moment in the cell cycle. The assumption is that the RNA polymerase works down the DNA strand and passes a particular site only once during each DNA replication cycle. There is even a case in which the sequence of synthesis of three enzymes is the same as that of their positions on the chromosome, but, as Mitchison points out, there are difficulties in that some organisms seem to have both types of control, and the way they are conceived of at the moment seems to suggest that they should be mutually exclusive. Clearly, this problem is not solved; all we can do now is glean some idea of the kinds of control that are possible.

This raises the question of whether or not DNA synthesis is necessary for the initiation of a new pattern of transcription. We do know that certain differentiations appear to be invariably preceded by cell divisions, and therefore an obligate relation is assumed. Furthermore, in the case of the differentiation in the mammary gland in the mouse and in a number of other systems,[8] it is known that any agent which blocks DNA replication blocks

139

differentiation. The only difficulty is that one cannot be certain that it is the transcription and not some other control point, such as translation, that is dependent on the DNA synthesis. Examples of the opposite situation are easier to prove. For instance, in the cellular slime molds Sussman and Newell[9] have excellent evidence that transcription of previously inactive genes can occur without cell division. Also Leach and Ashworth[10] have shown that the transcription of a number of enzymes can occur in either the G_1 or the G_2 phase of the cell cycle, again indicating that to a considerable degree the state of the DNA is independent of transcription. This being the case, one would want to examine further and with rigor those instances in which cell division appears to be absolutely necessary for differentiation and see exactly how they operate.

The implication from the enzymes that are produced continually during the cell cycle (linear enzymes) is that the amount of enzyme synthesis equals the amount of DNA that codes for it. It is doubtful whether this assumption is invariably safe; as one can see from Varner's table (Table 1), one could expect also changes in the rate of transcription.[11] In any event, it is obvious that one way to amplify a message is to have the gene repeated many times. This is exactly what has happened in the case of the DNA coding for a number of ribosomal RNA's. This is not the only method of amplification, but it is clearly an effective one.

It is impossible, when discussing control mechanisms of substance synthesis, to divorce transcription from the other control points. This must have already been clear in the preceding brief outline of transcription, and it will continue to be so as we look at translation and enzyme activity. It may be possible to have a better perspective of the magnitude of the problem when we attempt to look at the interconnections between the different control levels.

Translation

Translation involves reading off the *messenger* RNA in the ribosomes and, with the help of *transfer* RNA, producing poly-

140

peptide chains that correspond to the coded information in the *messenger* RNA. Control can be exerted in a number of ways, the time at which translation occurs, that is, the interval between transcription and translation, being the most obvious one. However, since this involves a timing control, it will be considered in the next chapter.

It is obvious that there are numerous ways in which the amount of any protein synthesized could be determined at the point of translation. In fact, it would be quite possible to have the cytoplasm loaded with specific *messenger* RNA's, which either are or are not used in translation. For instance, in dormant seeds there are large quantities of such messages which are not translated until water penetrates into the seed and in some way activates the translation mechanism.

A more important kind of control would be exercised if at the translation point there could be a significant alteration of the proteins produced, an alteration from the directions prescribed at transcription. The possibility that this might exist has been examined in detail by N. Sueoka and T. Kano-Sueoka.[12] They point out that the most likely way in which this might occur is the involvement of *transfer* RNA. There are multiple *transfer* RNA's for each amino acid, and one possibility, for instance, is that some of these may be degenerate or become modified in some manner so as to translate certain messages in different ways. It is known that different differentiated cells may have differences in their *transfer* RNA composition, and there are some instances in which it can be shown that these differences do affect the pattern of protein synthesis. But despite all this evidence, the Sueokas point out that the case is still not proved, but rather a likely and interesting possibility. If it is firmly established as a phenomenon, then one will want to know how common it is, and how it fits in with transcriptional control; they could both be operating simultaneously. Also one would want to know the mechanism of the *transfer* RNA modifications. Presumably this is also genetically controlled, but again it is clear that much work is needed.

Enzyme control

This is another vast subject which we shall summarize in the briefest and most superficial manner. Enzymes are exceptionally efficient and complex catalysts. Therefore one of the ways in which they control is by determining the rate of reactions, again a matter more pertinent to the discussion of timing than the production of substance.

Enzymes have many properties that reflect their control abilities. To begin with, they can exist in inactive and active forms. Varner has listed some of the possible ways in which the initiation of the activity can be controlled (Table 2), which again

Table 2. Possible points of control of enzyme action.
(From J. E. Varner, *Symposia of the Society for Experimental Biology*, 25, 198 (1971.)

Oxidation of sulfhydryl groups
Reduction of disulfide bonds
Hydroxylation of proline and lysine
Glycosylation of hydroxyl of amino acid side chains
Peptide bond cleavage
Phosphorylation
Acetylation
Adenylation
Removal of an inhibitor
Addition of an activator

emphasizes the extraordinary multiplicity of ways whereby any one step can be governed.

Besides being simply turned on and off, enzymes can undergo much more subtle regulatory changes. This has come from all the recent important work on allosteric enzymes. Enzymes can be affected by other substances, even their substrate of their product, or some precursor of the substrate, or some substance produced from the product after further enzyme actions; these possibilities will be considered when we talk about interconnections between controls. In fact, one enzyme can be affected simultaneously by an inhibitor and an activator that will balance each other's effects. As Changeux[13] has pointed out, there is a

142

close parallel here with what one finds in interneuronal connections: a cell body of a motor neuron will have synaptic attachments from other nerve cells and these peripheral cells can exert an inhibitory or a stimulatory effect on the central neuron's output of signals.

By such allosteric control mechanisms one has all the needed mechanism to account for feedback inhibition or feedback stimulation. All the kinds of elements that one finds in an electronic computer circuit are here: rectifiers, triodes, and so on. Furthermore, at any stage they can be influenced also by environmental cues; not all the signals are internally generated, as we shall see presently.

Interconnections between controls

Now that we have the elements, the parts necessary to make a complex web of interconnecting controls, we can begin to set up circuits, much as an electrical engineer can build a radio with his vacuum tubes or transistors. The difficulty is that the developing organism is such a maze of circuits (and they are not all painted in different colors or neatly printed on a circuit diagram) that to unravel them is exceedingly difficult. Here we shall adopt two approaches: first we shall briefly discuss some of the basic ways in which interconnections occur, a sort of taxonomy of biochemical circuitry. Then we shall look at what is known of various circuits at different levels, that is, the interconnections among nucleus, cytoplasm, developing groups of cells, developing social organisms, and finally organisms and the environment.

As far as a general model for circuits is concerned, we may begin with the three basic control points of transcription, translation, and enzyme activity. Next we must remember that all the molecular components of the cell, including nucleic acids and proteins, are made by a series of enzyme-promoted steps, so that the product of one reaction is the substrate of the next reaction. Often a reaction will bring two substrates together to form one product or make two products by cleaving one substrate. In this way the pathways of synthesis of different substrates will impinge on one another.

143

In classifying the interconnecting controls we shall follow Monod (Fig. 28). The best-known category is *feedback inhibition,* where the product of a reaction inhibits its further production. This may be an immediate effect on its own enzymes, or it may act on some earlier enzyme in the series, as was first demon-

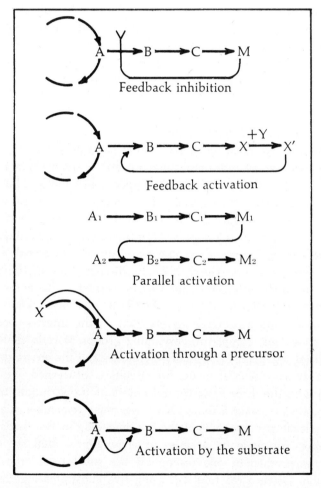

Fig. 28. Various "regulatory modes" assured by allosteric interactions. Arrows with solid lines symbolize reactions producing intermediate compounds (*A, B,* etc.), *M* represents the terminal metabolite, the conclusion of the sequence of reactions. Fine lines indicate the origin and point of application of a metabolite acting as an allosteric effector, the inhibitor or activator of a reaction. (After J. Monod, 1972).

strated by Gerhardt & Pardee[14] (and is shown in the top of Fig. 28). Instead of an inhibition, a product could stimulate a reaction to produce a rapid accumulation of a product. This *feedback activation* is known to occur in cases in which the product disintegrates and one of the products of the disintegration stimulates, thus maintaining the quantity of the terminal metabolite in a steady state. In *parallel activation,* the terminal metabolite of one sequence activates the beginning of another sequence. This is found in cases in which a number of metabolites are needed to accumulate together for assembly in a macromolecule. An enzyme can be *activated through a precursor.* A frequently observed example of this is the activation of the enzyme by the substrate itself. One might call these last examples *feed-forward activation;* it is obvious that one could not have feed-forward inhibition, for then the reaction would never occur.

Since we have the possibility of turning enzymes on or off or modulating their activity in either a positive or a negative direction, and since the product of one reaction may affect another, the possibilities for complex interconnections are enormous. This kind of network is the basis of steady states within organisms: it gives us the steady state of metabolism, and it can account for some of the changes found in development.

If one were to set up a theoretical model of such cascades of steps and make certain reasonable assumptions, one could easily produce a situation in which a small change in one substance (for example, an initial substrate) would produce a radically different end state.[15] It is possible to set up steady states in such a way that a slight push will put them into condition A or condition B (or C or D), and A and B are equally stable but are quite different in, for instance, the relative amounts of certain products. This difference in products could result in radical differences in the appearance of the cells in condition A as compared with condition B; it could be the cause of cell differentiation. The system is so simple that one may readily assume that many differentiations in development are exactly of this sort. The most obvious examples are from plants where there seem to be two possible states in which the organism can exist. For instance, in the leaf

145

of English ivy, the young plant has a different leaf shape from the adult (Fig. 29). If one propagates the juvenile tissue by cuttings, one can perpetuate the juvenile form, and the same is true for propagating the adult form with adult cuttings. Normally, the only way to have the adults produce the juvenile form is through the seed, although it is possible to cause reversal from the adult to the juvenile form by various experimental means, including the application of giberellic acid.[16] Furthermore, the tissue cultures from the two forms have different properties. The evidence that such differences come from steady-state shifts is from parallel work in the tissue culture of plant cells. Meins[17] has shown that the difference between two stable clones of tobacco teratoma cells appears to depend upon a difference in their metabolism. The two stable types are interconvertible and

Fig. 29. English ivy (*Hedera helix*), showing the lanceolate mature leaves and the characteristic star-shaped juvenile leaves. [From K. Goebel, *Organographie der Pflanzen* (Fischer, Jena, 1928), vol. 1, Fig. 529.]

the conversion can be stimulated by altering the key nutrients in the external medium. Another example is the well-known work of Skoog and Miller,[18] who treated tobacco callus tissue with different concentrations of auxin and cytokinin. Depending upon the relative amounts of these two substances, they produced root tissue, leaf and bud tissue, whole plant shoots, or callus. In all these cases one can assume that there is a network of reactions which are in constant turnover in the system, and that this network can be pushed to one or more stable states.

In the examples we have considered so far we have not specified whether or not they are steady states set up entirely by shifts in enzyme activities in the cytoplasm, or whether they could involve nuclear activities as well. In point of fact they could be either; they represent the general case. Let us now concentrate on a consideration of where in the organism the control mechanisms lie, and how they are interconnected.

Nucleus-cytoplasm and other levels of interaction
The usual situation is one in which both the nucleus and the cytoplasm are involved. If one looks at the feedback and feed-forward control loops in Fig. 28, it is obvious that these loops need not just involve cytoplasmic enzyme-promoted steps, but could reach back to the transcription and translation control points. There is no barrier between the two levels; quite the contrary, nucleus and cytoplasm are constantly interacting with each other. For instance, it is possible to give an example of alternative steady states that clearly involves the genome as well as the cytoplasm. This is the phenomenon of transdetermination of Hadorn.[19] The fates of the tissues of an insect imaginal disc remain constant despite repeated transfers in the abdomens of adult flies. However, occasionally a cell will switch and remain stable in a new direction (for instance, the insect shows antenna instead of leg differentiation), and this change clones consistently when implanted into larvae and sent through metamorphosis. But in this case we say that it is a steady state that involves the genome because the new condition involves the transcription and translation of a different set of genes (while our earlier exam-

147

ples could theoretically be accounted for by metabolic or cytoplasmic shifts from one equilibrium point to another).

There is now overwhelming evidence that the nuclei of the cells of multicellular organisms are, with very few exceptions, genetically equivalent. This view is a very ancient one, going back to T. H. Morgan, Hans Spemann, and others in the early part of this century. It was also realized that it was the differences in the cytoplasm that were responsible for the regional differentiations of an embryo. The basic idea was that nuclei surrounded by different cytoplasms would give forth different genetic information. The same notion is with us today, and in modern terms we imagine that the different cytoplasms in some way affect transcription (and possibly translation also) so that specific gene actions are called forth. The question of what localizes the cytoplasm in this fashion is not our concern at the moment (it will be examined in Chapter 8), but rather we want here to examine the question of specific synthesis of proteins regulated by the nuclear environment.

Examples of the immediate cytoplasm affecting gene action are legion. In classical embryology one sees examples among various sorts of mosaic development, where different parts of the egg have different cortical cytoplasms, and these differences ultimately result in the major tissue differentiations of the developing embryo. The evidence that the important elements reside in the cortex comes from the observation that centrifugation (except in some instances at very high speeds) will not alter these cytoplasmic effects. A beautiful example of this cytoplasmic effect is found in cortical differentiation of particular blastomeres in some invertebrate mosaic embryos. In the mollusc *Teredo*, certain blastomeres show cortical staining with silver impregnation that is lacking in the other blastomeres.[20]

More recent kinds of evidence come from work on nuclear transplantation and cell hybridization.[21] It can be shown by means of cytochemical methods that a dormant nucleus transplanted into a new cell active in RNA synthesis may swell and renew its own RNA synthesis, suggesting that the foreign cytoplasm has stimulated a new round of transcription. Goldstein

148

and Prescott[22] have shown in amoebae, with radioautograph and nuclear-transplantation techniques, that proteins enter (as well as leave) a nucleus, and it is presumed that this could be the way in which the cytoplasm gives specific orders as to what transcriptions are to be made. It is known that if isolated nuclei are examined for their immediate ability to transcribe certain RNA's, nuclei from different tissues show different capabilities. However, if one removes the protein from these nuclei, a much wider range of RNA's is produced; presumably all the DNA is available for transcription. This again returns to the idea that proteins in various forms, including histones, are responsible for repressing specific regions of the genome.

So far these examples of effects of the cytoplasm on the gene activity of the nucleus are confined to the immediate cytoplasm (or cell cortex) surrounding the nucleus. In other words, we have not strayed from a single cell. In multicellular organisms a new device appears: there are substances that are relatively mobile and can pass from one group of cells to another. These fall in to two main groups, although the distinction may not be as sharp as one imagines. The least mobile are the inductors, which characteristically do not wander great distances, but are macromolecules that pass from one cell to another provided the cells are in contact. It is obvious from the intensive work done on embryonic induction since its discovery by Spemann early in this century that there are many different inductors in the development of vertebrates and that some appear to have the characteristics of nucleic acids, some of nucleoproteins, and others of proteins.[23] It is not known in any instance how an inductor acts on a cell. It could either act directly on the cytoplasm and stimulate some specific differentiation or act on the nucleus and affect transcription. The latter hypothesis is favored, although the evidence is not strong. Of course, even a direct effect on the genome may involve numerous chemical steps, for it is possible that the receiving cell has special receptor molecules which in turn affect the genome.

The other kind of substance that stimulates differentiation is a hormone. These substances are far more mobile than inductors:

in animals they perfuse the entire body through the circulatory system; in plants, as we shall see later, they move in a directional fashion. This means that all the cells of an organism may be exposed to the hormone, but the fact that only some regions respond has to do with the localization of specific combining sites; as with inductors, there is a chemical reaction that takes place between the relatively small molecule of the hormone and some reactant.

There is evidence that in some cases the receptor molecule may be a protein and that the complex of the hormone and this specific protein has a direct effect on the genome. It is believed that steroid hormones, in general, are able to pass rapidly through the cell membrane and combine with a specific protein in the cytoplasm, and that the complex then passes through the nuclear membrane to the genome. An excellent example is provided by the recent work of Yamamoto and Alberts,[24] who give evidence that estrogen is very likely taken up by the cells of the uterus. In this way the estradiol-protein complex acquires an affinity for a specific region of the DNA in the nucleus, which, in some manner, ultimately stimulates the growth of the uterus. Another example is the recent work of Ohno,[25] who suggests that the male hormone, testosterone, binds to a protein that is repressing the regulatory genes on the chromosome which makes RNA messages that give rise to the male characteristics. In this case the hormone presumably is an inducer, in the sense of the microbial geneticist, by removing the repressor protein from the active site on the chromosomal DNA.

There are many more cases in which the combining site in the target cell is not in the genome, but either on the cell surface or in the cytoplasm. Consider, for instance, the case of the action of epinephrine on the liver cell; it serves as an excellent example of the general principles involved. The epinephrine (or adrenalin, if you prefer) wanders freely about in the entire circulatory system of the mammal, and in the liver it combines specifically with a particular kind of adenyl cyclase that is on the cell surface. This stimulates the cyclase to convert ATP into cyclic AMP inside the cell. The cyclic AMP activates an inactive kinase, an enzyme that in turn converts dephosphophosphorylase into phosphorylase. The latter enzyme has the specific ability of synthesizing glucose-1-phosphate from glycogen and inorganic phosphates. By two more enzyme-promoted steps the gulcose-1-phosphate is finally converted to glucose which again can circulate in the blood.

In this case the receptor molecule (adenyl cyclase) is on the cell surface. It is important to note here that other cells in the body have variant forms of adenyl cyclase that respond not to epinephrine but to other hormones. For example, the cells in the thyroid gland respond only to thyroid-stimulating hormone (TSH), and in this case the cyclic AMP does not stimulate the production of glucose, but of thyroxin. In each instance, after the original stimulus and production of cyclic AMP, there is a specific protein that binds inside the cell with the cyclic AMP, and the result is a series of enzyme-controlled steps leading, depending on the cell type, to the production of thyroxin or of glucose.

In this relation between cyclic AMP and hormones, there is an interesting evolutionary progression between prokaryotes and mammals. In the latter we have already seen that cyclic AMP plays a role that is intermediate between the hormone and the end product (for example, epinephrine and glucose). This is the reason why its discoverer, E. W. Sutherland, called it a second messenger.[26] In bacteria there is increasing evidence that it combines with a specific protein and that the combined molecules act on the chromosome so as to remove repression and permit the transcription of new, inducible enzymes.[27] Here the substance appears to be acting as a primary hormone, much in the way Ohno postulated for testosterone. If we compare the roles of cylic AMP in *Escherichia coli* and in mammals, it is clear that the substance has been prevented from acting directly on the genome by the intercalation of a large number of extra chemical reactions between the appearance of cyclic AMP and the end result. Glucose in the mammal is not produced by the cyclic AMP directly depressing the genome and thereby producing the necessary enzymes. The genome has already produced these enzymes independently of the cyclic AMP; the latter now acts by setting

151

these enzymatic steps in motion. In the beginning of the process there is also a difference. In mammals the presence of a hormone (for instance, epinephrine) is required to activate the cyclase; in bacteria it is the absence of food that provides the stimulus. Furthermore, the cyclic AMP is released outside the cells in bacteria, and in its presence other bacterial cells will start derepression and the production of new enzymes.

From an evolutionary point of view we can imagine the following sequence: cyclic AMP is a small, stable molecule, which therefore can diffuse effectively. By producing a protein that can combine with it specifically, it can perform a particular task, that is, provide an effective, localized trigger. In the beginning (prokaryotes) it will trigger the genome directly, but as more enzymes are held in the cytoplasm, it can easily transfer its effects there. If this particular small molecule has such virtues, it could also devise a way to be produced in a specific fashion by having its parent enzyme (adenyl cyclase) appear in different forms that respond to different hormones (which would now be the primary triggers). We end with one common link (cyclic AMP), which, through the different forms of adenyl cyclase, can respond to a variety of specific stimuli (hormones), and can in turn produce a variety of specific responses or end products. (All that is left unanswered is the spatial problem: why do some cells have one kind of cyclase and others another; why do some cells have a battery of enzymes that can be pushed by cyclic AMP into producing glucose, and others into producing thyroxin? We can never discuss the substances of development for long before arriving head-on at the question of their spacing.)

Before we leave the subject of hormone stimulation of differentiation, it should be pointed out that there are indications that some hormones may be acting directly on translation. This is from the work of J. Ilan[28] on the beetle *Tenebrio molitor* (a mealworm). He has evidence that when *transfer* RNA and enzymes from animals treated with juvenile hormone (which induces pupal molts) were mixed with messenger RNA and ribosomes from adult stages, they gave the particular tyrosine-leucine ratio characteristic of the molting pupa. The adult apparently makes a new *transfer* RNA synthetase which is not present in the juvenile stage, and the juvenile hormone blocks the synthesis of this adult synthetase. Of course, there are many changes in the formation of adult beetles, the increase in tyrosine production being only one.

If we look back at what we have said so far concerning the interconnections between control mechanisms, we can arrange these interconnections as ever-widening feedback or control loops. In prokaryotes the loops seem to be directly connected and intimately associated with the genome. As one moves up the scale to complex multicellular eukaryotes, the circles involve an increasing number of enzymatic steps in the cytoplasm; inductors make for larger interconnecting loops in that they link cells; and hormones, because of their ability to move faster and for greater distances than inductors, automatically increase the size of the control loops. With these systems we have some idea of the possible control mechanisms of substance production in single cells and in multicellular organisms; now we must look at control mechanisms in the differentiation of animal societies, specifically the organization of the division of labor in insect societies.

There was an ancient controversy over whether castes of either sex in social insects are directly determined by genetic differences in the individuals or by some environmental influence.[29] Today we know that, except for the stingless bees which do discriminate genetically between workers and the potential queens,[30] all other social insects so far examined have nutritional or chemical means of producing new castes. It is especially interesting that such chemical means have arisen independently in different groups. Honeybees have royal jelly, which somehow manages to turn any genetically female larva into a queen. In all social ants that have been studied, there is similar evidence for a nutritional influence on genetically identical individuals. In ants this nongenetic influence can, according to Wilson, operate in six different ways: (1) by larval nutrition (as in the honeybees), (2) by chilling the larvae, which increases the number of queens, (3) by altering the posthibernation temperature, (4) by the queen's giving off some sort of chemical inhibitor that prevents new queens from

153

forming, (5) by egg size, for the larger eggs tend to turn into queens, and finally (6) by the queen's age, for younger queens tend to produce a larger percentage of eggs that are likely to become workers. All these factors can be understood in chemical terms, in a change in either the nature or the amount of key substances.

But the matter can be seen much more clearly from the extensive work of Lüscher on caste determination in the lower termites.[31] First of all there is good (although incomplete) evidence that differentiation into a particular caste for any one individual is determined by internal hormone levels. Large doses of juvenile hormone cause the transformation into soldiers; smaller doses produce either a juvenile individual or a reproductive, depending upon the extent to which the nymph has already differentiated in one of these two directions. The reproductive condition is also favored by increased titers of ecdysone, the molting hormone. It is presently thought that all the forms or castes can be accounted for by the relative amounts of these two hormones.

The most important point, which Lüscher showed unequivocally, is that there are pheromones (the substances passed between individuals) that can affect the endocrine balance within an individual, which in turn affects its caste. If a king and a queen are kept in a small colony of *Kalotermes*, no secondary reproductives appear, but if they are removed, new reproductives appear at the next molt. In what is now a classic experiment, Lüscher separated with a wire screen a colony of worker nymphs from another colony with a king and a queen. The colony of workers produced secondary reproductives; they were no longer inhibited by the royal pair across the screen. But as soon as the new reproductives appeared they were eaten (and the whole process repeated itself). Apparently, because they touched antennae across the screen, the workers knew there were too many reproductives and immediately destroyed the newest pair. This could be prevented by a double screen; the antennae no longer touched and the two colonies lived in stability, each with its own reproductive pair.

The key to the inhibition of the development of new reproductives comes from the production of inhibitors by the existing reproductives. These are passed from one individual to another by licking. In fact, by attaching the termites head first in a screen Lüscher showed that the substances were passed in only one direction; if the individuals at the screen licked reproductives, they could pass on the inhibitor to those workers that licked their anal ends. The inhibitor would not pass in the reverse direction. It is, of course, a well-known fact that social insects in general, and these termites in particular, are forever licking one another, providing an effective means of pheromone transmission.

There is one final point in Lüscher's work that is pertinent here. The nymphs have successive molts, and if a nymph suddenly fails to receive the inhibition because of the loss of the reproductives, he or she will begin his or her development in the reproductive direction. All the nymphs will be affected in this way, but those that have just molted will be more "competent" and respond to the release from the inhibitor more readily than the older nymphs. The decrease in this competence resembles a random decay; it is an exponential decline. When the most competent nymphs finally emerge, they immediately begin to inhibit all further differentiation in that direction in the other workers.

It is clear from this best-understood case of caste determinations that we have added one additional feedback loop. The presence or absence of a pheromone, which is passed from one individaul to another by licking, is the most peripheral such loop. It directly affects the direction of differentiation of one or many insects. It does so by influencing the hormone balance, the relative concentrations of ecdysone and juvenile hormone inside an individual. How the inhibitor manages this is not known. We do not even know the chemical nature of the inhibitor. It is quite conceivable that the inhibitor has an effect on the behavior, on the brain, which is directly connected with the glands that secrete the hormones. The hormones affect the tissues, and they may do this directly on the nucleus, as has already been suggested for ecdysone, or on some cytoplasmic component (or both). The picture that emerges is that of a series of feed back or control loops that are connected one with another. At the top we would find pheromones that control the hormones which in turn control the cells, either the cytoplasm or the nucleus or both. There are chemical cycles within chemical cycles.

There is still one further factor which should be included here: That is the environment. The implication so far is that every step in the reproductive cycle is self-contained, but of course many of the steps are sensitive to the environment.

The environment can affect development in many ways. On a cell level, C. M. Child[32] pointed out that a heap of dissociated cells of a hydroid would form a new hydranth at their apex because the chemical environment at apex was quite different from that of the base: it had more oxygen and less CO_2, and any inhibiting substance could escape more readily. The pertinence of all the many instances of environment affecting development is more appropriately discussed in terms of timing mechanisms (Chapter 7) and localization mechanisms (Chapter 8). Here I merely want to point out that the synthesis of substances of developmental significance is often started or stopped by environmental cues.

To conclude this chapter it should be pointed out that we have never argued for any kind of ultimate control, because there is always a preceding step. If, for instance, one takes the position that sigma factors are the master controlling agents, one immediately demotes them from this exhalted position by asking what, in turn, decides which sigma factor is to act when. The answer to this problem again comes from the fact that here, as always, we are dealing with a series of life cycles in which the information developed in one cycle may be used in a subsequent cycle. A program for successive sigma factors has been built into the life cycles (which are the cell cycles for prokaryotes) and therefore can be prearranged in previous cycles for action in any one cycle. The ultimate control of the synthesis of substance—in fact, the ultimate control of all developmental events—comes from the power gained by repeated, successive life cycles. Only in this way can one achieve a vast array of complex, interacting events. Successive life cycles allow the accumulated information of millions of years to be used at a moment's notice. As far as control mechanisms go, there is no end: they go back to the beginning of life and are part of what we have called evolutionary development. The products of that control information are realized each life cycle.

Summary

The synthesis of substances in living organisms consists of a series of sequential chemical reactions. Ultimately the information necessary for all these reactions comes from the DNA, but this can be described only in terms of evolutionary development. In the development of a life cycle, there are a series of devices to bring forth and control the stored information. These involve transcription, translation, and enzyme activities. They appear in the form of controlled sequences or loops that can be wholly cytoplasmic, involve the genome and the immediate cytoplasm, or, in ever-widening loops, involve inductors, hormones in multicellular organisms, and pheromones in social insects. Many points in these controlled sequences are also influenced by the environment, so that the life cycle can accommodate to environmental changes.

7

Timing

As we have repeatedly emphasized, in development it is important to control not only the production of substances, but the time at which a substance is produced as well. Under the general umbrella of timing mechanisms we shall consider four aspects: (1) the control of rates of developmental processes, (2) the control of sequences of events occurring during the course of development, (3) the role of environmental cues in the timing of development, and (4) the control of the duration of life cycles. These four topics are different views of the temporal aspects of development.

The rates of processes

We are concerned here primarily with the rates of synthesis of substances. The speed of production can be influenced by a large number of factors. Consider, for instance, the rate of synthesis of a substance by an enzyme-controlled reaction. In the first place, the enzyme itself is a catalyst and therefore it has its own intrinsic rate at which it can turn out the product. If it is an allosteric enzyme, there are ways in which this basic rate of action of the enzyme molecule can be altered. Next, the temperature, pH, and other environmental conditions will affect the intrinsic rate. The amount of enzyme, the amount of substrate, and the availability of the substrate to the enzyme will all have a major effect on the overall rate of synthesis or accumulation of the product.

A good example that illustrates how one can vary the overall rate of product accumulation without affecting the reaction rates of the chemical processes comes from the work of H. G. Callan[1] on DNA duplication in eukaryotes. Using an ingenious technique, he showed that the rate of polymerization for each growing molecule of DNA is constant and roughly the same as in prokaryotes but the number of initiation points is far greater, so the amount of DNA synthesized per unit time is considerably increased.

Another way in which the overall rate of synthesis can be enormously influenced is by having one of the products in a chain of steps either disappear at a certain rate or remain inactive for specified lengths of time. Such, of course, is what one finds with *messenger* RNA. In bacteria the destruction of *messenger* RNA begins immediately after its formation; the decay follows first-order kinetics. The average life span of a number of *messenger* RNA's from *Escherichia coli* is about 2 minutes. This reflects the length of time they exist before they are degraded by enzymes. In eukaryotes there is also evidence of similarly short-lived *messenger* RNA's, but there is now increasing evidence that there must be many RNA messages which remain stable for long periods of time.[2] Furthermore, even among bacteria there is considerable variation for different *messenger* RNA's, and there is speculation that the rate of destruction of any particular *messenger* RNA may be genetically determined.

A few examples of evidence for relatively stable messages may be helpful. Information from experimental work on amphibians and sea urchins suggests that much of the protein synthesis that occurs between fertilization and gastrulation has been achieved by the translation of proteins from messages made before fertilization.[3] In the development of erythrocytes, RNA synthesis occurs at a relatively early stage and two or three cell divisions intervene before haemoglobin is synthesized; the *messenger* RNA is thought to last for about 2 days. At the other extreme there are cases of seeds that may survive in a dormant state for years, but as soon as water is added, they begin to synthesize protein. This occurs so quickly presumably because the *messenger* RNA has remained stable during the entire period of dormancy. A final, and especially interesting, example is the case of *Acetabularia*. In this alga, which totally lacks cross-walls, all the nuclei are condensed into one large body in the rhizoid.[4] This compound nucleus can be removed, and a nucleus of a foreign species substituted. Yet for a month or more, the proteins produced in the cap, which may be 5 centimeters away from the nucleus, are characteristic of the original nucleus. This fact is generally interpreted to mean that the *messenger* RNA is stable for that period of time.[5]

There has, of course, been considerable interest in the question of how the message is stored. In some cases it is assumed to be associated with the ribosomes in an inactive fashion; in other cases there is evidence for special proteins which combine with the *messenger* RNA's to form a stable nucleoprotein called an informasome.[6] It is quite possible that these nucleoproteins are polysome precursors which would bring the two ideas together. Clearly, this whole subject of message stability is one of continuing importance and active investigation.

In a general way this is roughly all one can say about the mechanism of governing the rates of process in development. There remains the important question of how such rate control might affect development. Long ago Richard Goldschmidt[7] proposed that genes acted by affecting the rates of processes. In this way he successfully accounted for all manner of differences in form and pattern of developing organisms. A good example was provided by E. B. Ford and J. S. Huxley,[8] who showed that the eye color of the shrimp *Gammarus* was the result of the speed at which the black pigment covered the red. Genes controlling the rates of synthesis of these two pigments would produce shrimp with either black or red eyes.

There are many other excellent examples, but it is clear that rates do not always play an important role, or at least not the kind of role Goldschmidt imagined. For instance, P. C. Newell and his collaborators[9] showed that if the cell mass of a cellular slime mold was disrupted and allowed to reorganize immediately afterward, certain enzymes that had already formed were resynthesized, but the second synthesis was achieved at a much faster rate. This clearly involved a new round of transcription as well as translation. As a result, the cell masses had twice the amount of each of the enzymes measured, yet development was normal. The pertinent point here is that there appears to be some unknown factor which normally governs the rate of enzyme synthesis, rather than the rate of enzyme synthesis being the limiting factor in controlling the timing of development. It is obvious that there are still important things to be learned about mechanisms of control of the rates of developmental processes.

Sequences of developmental events

The idea that reaction pathways occur in set sequences in living organisms is almost axiomatic. The synthesis of any substance within the organism involves a series of chemical steps, each one controlled by a specific enzyme. In this case, obviously, the product of one reaction is needed as the substrate for the next. Many of these chains of reactions, especially those associated with metabolism, involve cycles of such sequences, the Krebs cycle being perhaps the most famous.

Here we are concerned with the question of the degree to which chemical reactions involved in development are fixed in a set sequence. It is the same question asked by many developmental biologists back to the days of Aristotle; it is the basic concept of epigenesis. Aristotle[10] was most specific and said that if one set up condition *A* it would inevitably lead to condition *B*, which in turn would go on to *C*. Of course, he was not thinking in terms of chemical reactions, but of morphological states, for he and many epigeneticists who followed him were impressed by the fact that the egg or the early embryo has little resemblance to the adult (while the preformationists felt very differently and saw miniature adults in the germ). Epigenesis was the basis of Wilhelm Roux's *Entwicklungsmechanik*, the causal embryology of A. Dalcq,[11] and the epigenetic landscapes of C. H. Waddington.[12]

161

To put this aspect of development on a biochemical level, let us examine the evidence for sequential transcription and translation of proteins.

The evidence that genes act at different times during a life cycle has been known for a long time. In 1932 J. B. S. Haldane[13] wrote a splendid paper giving numerous examples. Genes act not only during different periods of early development, but also in the adult. Of greatest interest to us, perhaps, are the old experiments on the genetics of coiling in the snail *Limnaea*.[14] The coiling may be either dextral or sinistral, and the direction is determined at a single gene locus. However, the offspring always coil the same way as their mother, irrespective of their own genetic constitution. The explanation is very simple: the gene expresses itself in the next generation, that is, in the germ cells (oöcytes) of the adult and for this reason it is first evident in the direction of the spiral cleavage, which is an early reflection of the direction of the ultimate coiling of the shell. H. Spurway[15] has a similar example in which a gene expresses itself by producing no grandchildren. In this instance the germ cells of the next generation are sterile, so again a gene action is spanning an entire life cycle.

The beauty of these last two examples is that they show that even on the gene-action level all the information does not issue forth in one life cycle. It has been one of the central themes of this book that the cell is the reservoir of previously expressed DNA information, and that each life cycle which is derived from the cell makes use of information directly from the DNA and that which is secondarily derived from the DNA in the form of various substances and structures in the cell. By the cell's being the minimum of expressed information, elaborate life cycles are made possible. Now we are saying that there can even be a delay of what one assumes to be transcription and translation that can span a life cycle.

Since we are concerned with timing mechanisms here, the key point is that transcriptions do not all occur at once during development, but in a definite sequence. One of the clearest examples of such successive transcriptions is in the sporulation of bacteria.[16] Not only do a number of enzymes appear at the successive

162

stages of spore formation, but, from the use of actinomycin D as an RNA-synthesis inhibitor, the evidence is convincing that in each case the production of the enzyme is preceded by a round of RNA synthesis. What is particularly important in this example is that Mandlestam and his co-workers have shown that any mutant that lacks ability to synthesize any one of the enzymes in the sequence stops its development at that point. Therefore it is not merely a sequence, but a causal sequence, in that the activity of one enzyme depends upon the prior activity of an enzyme transcribed previously; it is a locked-in sequence of events.

Sequences have been shown by Sussman and his associates[17] in the eukaryotic cellular slime molds. They have determined the specific activities of a number of different enzymes and have shown that each one peaks at a characteristic time during development. Newell[18] points out that not all the enzymes are necessarily required for the development of the fruiting body, although clearly some are. The evidence that the sequence is causal (that is, that the synthesis of one enzyme depends upon the synthesis of a previous one) is not so clear-cut as it is in spore-forming bacteria. In the mutant studies there are no convincing cases of series in which if one enzyme is blocked the subsequent ones fail. However, there is good evidence that the order of their appearance is consistent. A number of mutants have been discovered that develop at different rates, but the sequence of enzymes is the same regardless of the overall rate of development. Also, as we pointed out previously, Newell showed that disaggregated cells upon reuniting repeated a round of synthesis of a series of enzymes, and again the sequence remained the same.

Another example comes from studies on the cell cycle which have been admirably reviewed by Mitchison.[19] There are two reasons why this example is especially useful to us. The first has already been discussed: there are two ways in which successive transcriptions are thought to be regulated. In a few cases the sequence of enzyme synthesis is the same as the sequence of genes on the chromosomes, and therefore a linear reading of the chromosome would account for the sequence. However, in the usual situation no such relation can be found. Here it is assumed that the activity of one enzyme synthesis is causally connected to that of the next by an oscillatory feedback mechanism ("oscillatory repression"). But besides this insight into possible mechanisms of sequential transmission, there is also the very interesting evidence marshaled by Mitchison that the cell cycle is in fact made up of two quite independent sequential cycles. One he calls the "DNA-division" cycle and the other the "growth" cycle. The former is concerned entirely with all the events in the duplication and separation of the DNA, and the latter with the synthesis of enzymes and products connected with the increase in cytoplasm. The evidence for these separate cycles comes largely from the fact that it is possible to block the DNA-division cycle for brief periods and not affect the growth cycle at all. There are then at least two sequential series of transcriptions that are occurring independently of one another, although they presumably are normally synchronized to some extent.

These kinds of sequences, which are set in self-contained units, have already been discussed in Chapter 4, where we referred to them as superunits of gene action. As was pointed out there, these closely coordinated sets of gene actions are examples of what de Beer calls heterochrony; different structures in the embryo can appear at different times relative to one another. This is the basis of neoteny, in which the appearance of mature germ cells can change in time in relation to the maturing of the somatic structures of the organism. In this way normally immature larvae of some animals (such as the amphibian axolotl) became sexually mature. Previously we stressed why natural selection would favor such a system; here we point out that set sequences of developmental steps often come bound in packages, and, although within each package there may be a rigid epigenetic sequence, the sequence of appearance of the packages may be and is changed during the course of the life cycle. It is impossible to guess how many such packages might exist or what portion of the gene actions is parcelled in this way, but clearly the phenomenon is relatively common. It may well be that all life cycles are made up of such sets of sequences, each one of which varies in its degree of internal rigidity and its ability to exhibit heterochrony.

A multicellular organism is a collection of cells, and a social organism is a collection of multicellular individuals. It is possible for different identical units within a whole to undergo a set sequence leading to differentiation at different times. The result will be that the whole organism or the whole colony will have an entirely different character if the differentiation of parts is staggered in this way. For example, the workers in a beehive go through their developmental cycle of activity at different times (depending on when they emerge), with the result that at any one time there are workers of all ages, and therefore all the tasks of the hive can be carried out continuously.[20] Since such temporal differentiation involves an integration of the whole organism or colony, it will be considered in more detail in the next chapter.

Environmental cues

Thus far in the discussion there has been the implication that every step in the reproductive cycle is self-contained, but of course this is not the case. There are many ways in which the inner chemical events are kept in harmony with the environment. Let us begin with the simplest cases and work toward the more complex ones.

The development of a bacterium is utterly at the mercy of the environment: if the weather is warm and there is ample food and moisture, the organism flourishes and rapidly goes through many life cycles; if a drought or a freeze occurs, all activity ceases. One dodge to survive in adverse conditions is the formation of resistant spores. Most spore-forming microorganisms will, upon depletion of the food supply, begin the process. The stimulus for the formation of the resistant stage is some chemical cue. In some forms, such as cellular slime molds, there is evidence that a drop in humidity will also hasten the process of spore differentiation.[21] Once the resistant spores are formed, they will remain dormant until favorable growth conditions return. The appropriate temperature range is a requirement for the germination of the spores or cysts of all microorganisms. In some, besides temperature, sufficient humidity is the only other requirement; in others, germination is greatly speeded up by the presence of some metabolite in the substratum.

Higher organisms also go through periods of quiescence, for

instance, the diapause of an insect or the seed stage of an angio-sperm. In these cases what was said of microorganisms generally applies, but there are more complicated mechanisms as well. Many seeds and insect pupae require, for example, a period of chilling (vernalization) before the dormancy can be broken. There are instances in plants and insects where light is needed; certain seeds will germinate in the presence of red light. This is because the seeds contain phytochrome, which absorbs the light and di-rectly stimulates the germination process.[22] It is uncertain how the phytochrome manages this, but there is some evidence that it alters the permeability of the cell membranes, and in this way it might ultimately initiate the intake of moisture from the out-side. By having a light-sensitive pigment the seeds can be stimu-lated to resume their cycle.

Phytochrome is a good example because it is also known to be involved in numerous other processes, including the initiation of flowering and seed formation. Plants may be sensitive to the lengths of night and day: long-day plants bloom in the spring, when the days are lengthening, and short-day plants in the fall. The phytochrome system is used as a measure of the length of darkness and is responsible for setting the flowering process in motion. Insects are also sensitive to the length of the day, although they use a different light-receptor pigment.[23]

These methods of measuring the seasons are thought to be connected with biological clocks; in some way the organism com-pares the timing of external events with its internal pacemaker. The whole question of endogenous biological rhythms has be-come of increasing interest because it is apparent that organisms in general, even single-celled organisms, can keep time with a high degree of accuracy. They can, in the absence of external in-fluences, mark off a 24-hour period (or something consistently close to 24 hours), a phenomenon known as a "free-running" circadian rhythm. These rhythms are especially sensitive to, and can be reset by, light changes, so that life-cycle events will har-monize with environmental events.

So far, the only role that biological clocks are known to have in development is in the examples already given, in which an orga-

166

nism keeps its life history in tune with external daily or seasonal changes. The crucial question is whether or not the endogenous, free-running rhythm might not also be controlling other internal developmental steps. One of the most prevalent theories of how the organism keeps time involves internal oscillators. Oscillations can occur in various chemical systems, such as enzyme reactions in which there is a rhythmic overshoot of substrate and then available enzyme. We saw such an example in the "oscillatory-repression" model to explain the stepwise production of substance during cell division. But there is no evidence to suggest that these particular oscillations are the ones that constitute the free-running biological clock.

In the absence of any firm facts, there is ample opportunity to makes models in which endogenous rhythms play a role in development. Goodwin and Cohen[24] have produced an ingenious phase-shift model of oscillators which they suggest could account for all sorts of "positional information" in the embryo, but unfortunately the scheme remains hypothetical.

The timing of the life span

We have already referred to the fact that one of the requirements of natural selection is the elimination of individuals, and therefore one might expect a selection for the probability that organisms shall have a finite life span. In some instances the harshness of the environment seems to automatically take care of the situation; all microorganisms, for example, seem largely to be eliminated by the severe changes in the external world. Even among passerine birds in temperate climates, it is estimated that on the average about 50 percent return to the same nesting area each year; this should be compared with a minimum 89 percent return in the more benign tropics.[25]

Here we are concerned with those cases, such as the tropical birds, in which the external world does not do the job and the organism provides a built-in mechanism of self-destruction. This has been called senescence or aging. The idea that the degeneration of the soma begins some time after the peak of the reproductive period and is a genetically determined phase of the life cycle is fairly universally accepted. The difficult problem is to understand how the genes manage to make the body senesce. This is compounded by the fact that one finds senescence in such structurally and physiologically different organisms as, for instance, mammals and angiosperms. However, this appears most reasonable from the evolutionary point of view. All that is required for natural selection is a degeneration of the organism's soma, and there are obviously innumerable ways in which this could be achieved mechanically. Therefore, gene mutations that increase the chances of death by internal disorder at a particular point in the life history will be under positive selection. These may involve tissue changes, organ malfunctions, or higher susceptibility to a particular disease; the list could no doubt be made very long. And, as we have repeatedly stressed, not only is convergence expected, but any one kind of organism may have many mechanisms that favor an increase in decrepitude at a certain stage. Each new mutation that fosters such mechanisms will be selected for (or left unaffected), but will certainly not be selected against.

In each of the kinds of timing we have discussed in this chapter, namely, rates of processes, time sequences, susceptibility to external cues, and the timing of the life span, there has been the clear implication that there is genetic control. In some instances one assumes that there is a very short route between the message that lies in the DNA and the end result. For example, the rate at which a particular substance is made may depend on a single enzyme that is directly transcribed and translated from the DNA. In others, the route may be exceedingly indirect and circuitous. It could depend upon the presence of one or more purely extraneous substances that in some way influence the activity of key enzymes. Or it might be by controlling the time between transcription and translation, which could be managed in a number of different ways. Even spatial considerations could be involved. For instance, one of the steps might require diffusion and therefore be limited by the rate of diffusion and the distance required to diffuse. This again illustrates the impossibility of completely separating (except in abstraction) the processes of producing, timing, and placing substances.

The sequence of transcriptions appears to be fixed in numerous instances. One might imagine that this occurs by a feedback mechanism within the cytoplasm. The transcription of protein *B* might not be able to occur unless protein *A* has been transcribed previously. Protein *A* might wander back directly into the nucleus and in some way permit the transcription of *B* to occur (perhaps by affecting the sigma factor), or there might be innumerable chemical steps in between in a very large feedback loop. Since the feedback loop itself is capable of marking a set amount of time, the interval between one transcription and the next can be precisely controlled.

It is harder to imagine how timing occurred over longer time spans, how, for instance, a gene acts in a postreproductive period of an organism with a long life cycle. There is, however, no theoretical difficulty; the mechanisms we have postulated could apply equally well to long time periods. The number of steps could be increased and the rates of various steps decreased, so there is little restriction on how long a DNA-generated time cycle might last.

A number of interesting studies have been made on the genetic control of biological clocks. These have recently been reviewed by V. G. Bruce;[26] genetic-clock variants have been shown to exist in bean plants, *Neurospora, Paramecium, Drosophila,* and *Chlamydomonas* in Bruce's own fine study. He points out that there are two major kinds of mutants: ones in which the free-running rhythm varies, and ones in which the period of the clock itself remains fixed but the overt expression of the clock, such as eclosion time in *Drosophila,* can vary in its phase relation with the basic clock, as was demonstrated by Pittendrigh.[27] In *Chlamydomonas,* Bruce has been able to show the existence of both kinds of mutants: ones with different fundamental periods and one that was a phase mutant. Clearly, by this kind of genetic analysis it may be possible to obtain some insight into the component parts of the clock. This, along with a molecular analysis, may ultimately uncover the mechanism of the clock and at the same time reveal to what extent biological clocks play a role in the timing of developmental processes.

Summary

The information contained in the DNA is sufficient to time the various events in the developmental life cycle. It does this in an enormous variety of different ways, which can be roughly grouped into control of developmental processes by affecting their rate, their duration, and their sequence in time. Since every chemical reaction in the developing organism has certain temporal properties, and since these will in turn affect the temporal properties of subsequent related reactions, one may think of the timing of all development involving all sorts of modifying feedback loops, much as was described for the synthesis of substances.

It has also been suggested that many of these reactions may oscillate, and a number of such oscillators may have further powers of time control, including the production of the internal or endogenous rhythms that are characteristic of so many organisms. Furthermore, at all stages the timing of developmental events can be influenced by the environment, either by producing harsh or favorable growth conditions, or by careful synchronous control to ensure that developmental events are in step with diurnal or seasonal climatic changes. The most sophisticated way in which the development is harmonized with the outside world is the gauging of night and day and of the lengthening days of spring and the shortening days of fall by means of the internal circadian clock within the organism.

8

The Localization of Substances

As is so often true in any kind of analysis, there is something artificial and abstract in separating the process of synthesizing a substance and its timing from the problem of where in the developing organism the substance is produced. It conjures up the old problem of the physiologist of 50 years ago, who worried about the absurdity of separating structure from function. In both cases there are virtues and dangers in such an analytical separation. The real danger comes from forgetting totally the other part of the equation, and this can be easily avoided. The advantage comes from being able to penetrate deeply into the problem and doing it in such a way that it is illuminated and clarified.

There have been few attempts to bring together the biochemical information of development with pattern information. This is because we are only at the beginning of the science of development, at least in terms of important progress. Recently the biochemist has made great advances which have been very briefly summarized in the previous two chapters. The consciousness that there is a spatial problem has been largely expressed by the developmental biologists, and the appreciation of it goes well back into the last century. Hans Driesch,[1] for instance, did a beautiful series of experiments in which he showed that there was a relation between the size of the regenerated hydranth in the hydroid *Tubularia* and the size of the piece of stem that bore the new hydranth. He understood that there was some way in which a portion of excised tissue could give a proportionate development; some of the tissue remained stem tissue, and the rest became hydranth tissue. Since that early work, this has remained the central problem of embryology and developmental biology. Lewis Wolpert[2] has again brought it to people's attention by saying that in development there must be "positional information," and he effectively simplifies the basic question in his "French-flag problem," which asks what mechanisms could account for the three equal stripes, no matter what the overall size of the flag. This is the identical question of regulation asked by Driesch and so many others for almost the last hundred years. It was Driesch who, as a result of his studies on regulation, suggested that the fate of a cell is a function of its position. He cut a sea-urchin embryo down the animal-vegetal axis and instead of getting two half embryos, as he expected, he produced two perfect dwarf pluteus larvae, a result that sustained his dictum. In the early part of this century others, especially T. H. Morgan,[3] pursued this approach to the pattern of development and made extensive speculations on how development and regeneration might be controlled.

E. G. Conklin and E. B. Wilson discovered that some invertebrates did not "regulate" as Driesch had shown, but were determinate. If a part of such a mosaic embryo was removed, the embryo produced not a smaller perfect embryo, but one that lacked the removed part. At the time it was thought that this involved an entirely separate kind of development, but later the two schools were unified. Most developing organisms, as they approach maturity, become relatively fixed or determinate in their differentiation. Usually tissues go from a state in which they can regulate to one in which they are determined or fixed. In mosaic embryos this point comes very early in development; in regulative embryos it occurs late.

In one way or another the concern with how pattern was established has been the central theme of developmental biology. For

172

instance, when Spemann first discovered induction, he called the dorsal lip of the blastopore the "organizer" because in the beginning he thought that he had found the pattern-producing region; only later was it clear that induction is simply a stimulus to a potential pattern already there.[4] There was a spasm of hope when the term "embryonic field" of Gurwitsch[5] appeared, because this seemed to underscore the nub of the problem; for a long time it served as a useful stimulus for biologists to look in the right place, but it unfortunately did not solve the problem.

The most ambitious attempt to solve it was that of C. M. Child[6] in his theory of metabolic gradients. He alone offered a simple and straightforward solution; his substantial contribution has not had the appreciation it deserves.

Interest in pattern formation has never flagged among embryologists. Furthermore it has been one of the prime concerns of geneticists; there is a large literature of pattern-influencing genes. It has also been an important concern of theoretical biologists and applied mathematicians. The pioneer was Rashevsky (although for various reasons his work is largely overlooked nowadays), and following him, Turing.[7] Today there are many people working on this aspect of development and producing a variety of interesting models. In fact, the difficulty in which we find ourselves at the moment is that there are so many theories, so many facts, and so many organisms that develop that it is only too easy to become fogged in. To avoid this state we must now take a systematic look at how substances are localized.

The easiest way to do this is to think of the placement mechanisms themselves. We shall begin, in this chapter, with simple physical forces such as the bond interactions between molecules, then consider diffusion, radiant energy, and electrical energy, and go on to the kinds of movements of substance that require living motors to do the moving. Since our ever-present interest is in control mechanisms, we want to understand how the pattern of distribution of substances is consistent from one life cycle to the next, a subject that will be introduced in this chapter but will be the main concern of the next: the grand question of pattern control and proportional development.

173

The localized appearance of a substance can occur in two ways: (*a*) by the synthesis of the substance in one spot or (*b*) by the movement of the substance, by either physical or "biological" means. Consistently with all aspects of development, generally both processes occur together or alternately; they are convergent and not mutually exclusive. Let us now examine the various ways in which this localization can be brought about.

Molecular bonding

If molecules are brought together by the forces of attraction between them to form a symmetrical structure such as a crystal, this constitutes localization. In this case it is localization by chemical bonding. Obviously, the distances over which these forces act are minute; nevertheless, they do involve a positioning of the "unit cells" of the crystal in a beautiful, regular pattern.

Today no one needs convincing that the forces of attraction between molecules and the configurations that result are important to biology. The cornerstone of modern molecular biology is the crystallographic study of DNA; from elucidation of the structure of DNA, its remarkable properties of template formation automatically followed. Understanding of DNA came from its native three-dimensional structure. The mineralogist and classical crystallographer know well the importance of how molecules pack, and how from this packing certain external shapes or symmetries emerge. One of the most beautiful and simple laws of nature is Haüy's law of rational indices. It says that only certain symmetries are possible and the reason for this is that there is only a limited number of ways in which identical units or molecules can be stacked. Molecules stack in perfect rows with identical orientation; they do so because of the attractive forces between them and between the atoms, which are arranged in identical fashion within each molecule. If the molecule is asymmetrical, the crystal may have a polarity; one face may, for instance, have different piezoelectric properties from another, or they may differ in their susceptibility to some solvent.

As the size of the molecule increases, the crystallographic problems change.[8] In very long molecules, for instance, especially in

174

ones that are likely to be contorted, there is increasing difficulty in achieving a regular stacking. Also there are situations in which substances with large molecules no longer exist in a simple concentrated solution. The solution will be dilute, and there will be an aggregation of macromolecules because of a bonding by specific chemical groups that lie in a certain location on the molecule. As a result we no longer find the perfect and restricted symmetries of supersaturated solutions of smaller molecules; a whole new range of geometries is possible which bear no relation to the law of rational indices (Fig. 30). But the important thing to remember is that in both cases, large and small molecular aggregates, the forces that hold them together are the standard forces of attraction between atoms and molecules. The new geometries are entirely the result of the increased size of the molecules and their concomitant increased complexity.

For clarification it should be stressed that the bonds which hold macromolecules together are generally known as "weak" interactions. They may involve attraction or repulsion. They include van der Waal's interactions, hydrogen bonds, hydrophobic bonds,

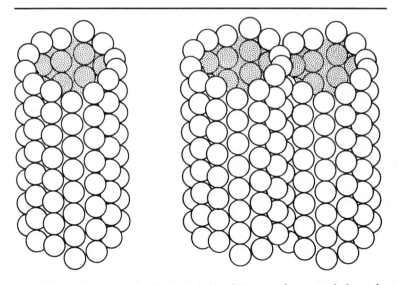

Fig. 30 The fine structure of microtubules: *left*, a single microtubule; *right*, a double tubule. (After D. L. Ringo and D. Chasey.)

175

and ionic bonds. Some molecules are asymmetrically charged or nonpolar, and this strongly affects their grouping behavior. This is especially true because nonpolar molecules are relatively insoluable in water. These bonds are called weak in that they can be disturbed easily by increasing the thermal energy, yet clearly they are sufficiently stable to preserve the form and the inner construction of living organisms. Very large, long molecules such as proteins will be affected by the bonding to produce different levels of structure. The primary structure is the sequence of amino acids on a backbone. These can form a secondary structure by such chains lying side by side being held together by hydrogen bonds. The result may be sheets of polypeptide chains (as in silk) or, more commonly, a helical structure. These helical aggregates of chains can in turn bend and twist to form a complex tertiary structure that is held together by disulfide bonds, bridges between the sulfur on the amino acid cysteine which may appear at intervals on the backbone. Proteins can bind to other substances: nucleic acids, lipids, and a whole variety of special groups of lower molecular weight.

These complexes, that is, the protein with or without the extra substances, can come together to form crystallike aggregates which will vary enormously in the perfection of their external symmetry. In some viruses there will be beautifully regular structures that seem to conform to the crystallographic rules of simpler molecules, while others will come together in all sorts of shapes that are quite impossible from a crystallographic point of view. The most common form of assembly is a simple membrane, which is found in all cells. Tubes are also common, the microtubules of flagella being the most obvious. There are many filaments; in fact, muscle is largely made up of masses of such filaments lying in parallel. Globular structures are found in the form of ribosomes. It is perhaps not surprising that such variation should exist when one considers the heterogeneity of the fine structure of the macromolecules, most especially the proteins that supply the basic material for all these structures.

One of the most beautiful examples of self-assembly in macromolecules is the formation of the T_4 phage. From the elegant work of Wood and Edgar and others[9] it is known that many of the parts of this complex phage are made up or assembled separately, and then the completed subunits are put together to form the final phage (Fig. 31). Initially it was assumed that the subunits were made entirely by self-assembly: the macromolecules had certain bonding capabilities that were totally responsible for their coming together. Furthermore, the assembly of the subunits into the form of the main structure was thought to occur in much the same way. However, recent work of Laemmli and King[10] has clearly demonstrated that some of these steps are possibly not self-assembly in the strict sense; they involve the action of special proteins. For instance, there are enzymes that cleave and alter the structural proteins during the assembly of the head, and there is a special protein needed for the transformation of coiled proteins into straight fibers prior to the assembly of the tail fibers and the collar. The development of T_4 phage is a combination of self-assembly and gene-controlled steps; we might call it a "quasi-crystalline" state.

Aggregates of macromolecules within cells come together in

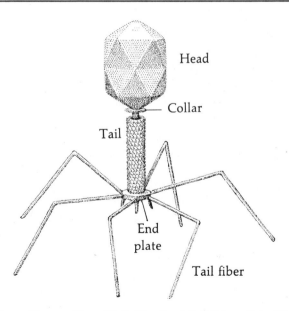

Fig. 31. Intact T_4 virus. [From W. B. Wood and R. S. Edgar, *Scientific American* (July 1967).]

interesting combinations to form organelles, structures that are of a higher level of organization. The microtubules in cilia are in a specific pattern of nine peripheral pairs and a central pair, and between the tubules there are regularly placed and oriented structures. The flagellum is attached to a basal body that also has an organized and fixed structure (Fig. 7). The mitochondrion is largely composed of membrane that forms convolutions into flat cristae (Fig. 6), and we could go on with the list. In each of these organelles there is a great mixture of different kinds of macromolecules which are combined to form the structures so that it is not a homogeneous aggregate, but a highly heterogeneous one. From what we know of phage assembly we could assume that part of the form of the organelle is due to the weak interactions between the various macromolecules, especially the proteins, and part due to the action of special proteins (often enzymes) which, in some way, are required to put the structure together. Therefore they might also fall into the very general category of quasicrystals.

Special interest may be attached to the way in which organelles grow and divide. The mitochondrion does so by merely pinching in two and increasing the amount of membrane. The flagellum cannot divide at all, but is in some way built by the basal body. The basal body can replicate and it does so in a most peculiar way. The parent basal body (or a centriole) looks like a short cylinder of *millefleur* glass. It has nine truncated microtubules in a circle but no central tubules, unlike a flagellum. If the basal body is to sprout a flagellum, the cylinder of the basal body forms a continuum with the cylinder of the flagellum. If the basal body divides, the parent structure remains intact, but the new basal body forms some distance away and at right angles (Fig. 32). There is no parallel in crystallography or even in the quasicrystallography of macromolecules; it is a most unusual phenomenon. The situation is further confused by the fact that there is evidence that this is not the only way new basal bodies appear: they can also form *de novo*, or form in large groups or clusters.[11]

The next level of structure is that of a group of organelles which are in some way fixed in a larger unit. The obvious example is the cortex of ciliates, in which, as in *Paramecium*, for instance, there are rows of basal bodies surrounded by a repetitive complex of structures (Fig. 8). In these cases the whole surface of the organism appears to take on the very same quasicrystalline properties that we have described for individual organelles. This is manifest in a number of ways. In cell division in *Paramecium* there is a doubling of many of the basal bodies that lie in rows along the longitudinal axis of the individual. When division occurs across the waist of the original cell, the rows of basal bodies or keneties remain intact, but since the basal bodies themselves are multiplying, soon the original number in each kinety is restored so that both daughter cells resemble their parent. In this case there is no total reconstruction of the cortex, but in each cell division the existing cortex replaces the lost parts. It has been shown in different ciliates, including *Paramecium*, that if an anomalous strain exists with an extra kinety (or a row lacking), the daughter cells directly inherit the anomaly.[12] The integrity of the cortical structure is passed on directly in reproduction; it is not recycled and reformed *de novo*. Therefore in this sense it behaves like a giant macromolecular aggregate that is capable of growth but not of reorganization. Any variation, such as an extra row of basal bodies, will be passed on directly at each cell division.

These same properties are shown in other ways which depend upon the fact that the cortex is polar. There is a strict anteropos-

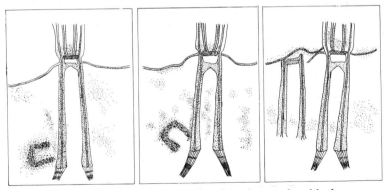

Fig. 32. The origin of a new basal body: (*left*) the new basal body appears as a dense area at an angle to the parent basal body; (*middle* and *right*) the new basal body grows and assumes a position parallel to the original basal body. [Original drawings from electron micrographs of R. V. Dippell in T. M. Sonneborn, *Proceedings of the Royal Society of London, B, 176*, plate 29 (1970).]

terior directionality to the cortex, which can be readily seen at the fine-structure level (Fig. 8). Each kinety is made up of basal bodies to which are attached strands (kinetodesmies) that are tapered. These all lie to one side of the basal bodies, and their tapered ends all point in one direction. The subunits of the macromolecular structure of the cortex are visibly polar. Fauré-Fremiet[13] discovered that he could produce cell fusions in the ciliate *Urostyla* by putting it for short periods in a dilute solution of formaldehyde. If the cells fused so that their anteroposterior axes were in the same direction, they became stable doublets or triplets that went through numerous divisions in this state. If their polarities were not aligned, they became unstable and one of the cells was absorbed into the other. The whole surface has certain steady-state configurations.

Recently Sonneborn and his co-workers[14] have done an interesting experiment in which, by microsurgery, they reversed a small portion of a kinety. The result is a *Paramecium* with part of one row of basal bodies in which the fine structure and details are all pointed 180° away from the rest of the surface. This anomalous kinety is now inherited; it appears to be a permanent fixture of the progeny (which have been carried through 800 generations). This again shows that the cortex is made up of macromolecules that assemble in a particular pattern, and that this pattern, even in a disturbed state, is directly inherited. As Sonneborn points out, this is nongenic inheritance, yet clearly the synthesis of the substances necessary for this replication of the surface macromolecules must be controlled by the genome. It is an exaggerated and beautiful case of what we have emphasized before: direct inheritance from one generation to the next is not restricted to the DNA of the genome, but many other substances and structures are built up from previous cell cycles. In this case we have a large and exceedingly complex cortex whose pattern of fitting together is a property of the macromolecules at the cortex and is not directly under nuclear control. Over what must have been a long time and a vast number of cell cycles, a surface structure evolved. The structure itself had properties such that its immediate form is independent of the nucleus; at the same time, it is totally dependent on the nucleus, we presume, for the synthesis of its specifically shaped building blocks.

The fact that a gene change can affect form by self-assembly is well illustrated in the case of sickle-cell anemia. This disease is caused by a mutation that alters a single amino acid in the haemoglobin molecule, and as a result all the red blood cells of individuals homozygous for this gene have an abnormal shape. One can imagine that many of the differences between species of organisms with a complex cortex, such as ciliates, could be accounted for partly on this basis. One must also leave room for the possibility that there might be special proteins which, as in the building of the T_4 phage, contribute in various ways to the macromolecular construction. This may also be true of various other shelled unicellular organisms which could have quasicrystalline forces involved in their shaping. The radiolaria, the foraminifera, the diatoms, and the dinoflagellates have an enormous variety of cell shapes, as Ernst Haeckel in the last century and D'Arcy Thompson in this century have emphasized.[15] This variety could indeed be interpreted in terms of genetic changes that alter the quasicrystallinity, either by altering the shape of the building blocks themselves or by changing the modifying proteins that affect their assembly.

There is an additional feature of the ciliate cortex that is of great importance: curious organized movements in the cortex play a vital role in certain developmental phases. A striking example may be seen in the normal cell division of a species of *Halteria* discovered by Fauré-Fremiet.[16] If the stages of division are followed from silver impregnations, it can be seen that the new mouth region forms at the posterior pole and the teardrop-shaped rows of bristle basal bodies divide longitudinally in the equatorial region (Fig. 33). Should this cell row cleave across the middle in the conventional way, then obviously the bristle rows would be pointing in the opposite direction in the two daughter cells. This does not happen, as is clear from Fig. 33, and a new anterior end appears as the new mouth moves to one side. The lower set of bristle rows moves toward the new anterior end and the two daughter cells end up about at right angles to each other at cleavage.

These are movements of structures over the surface, a phenomenon that cannot be easily understood. It is almost as though

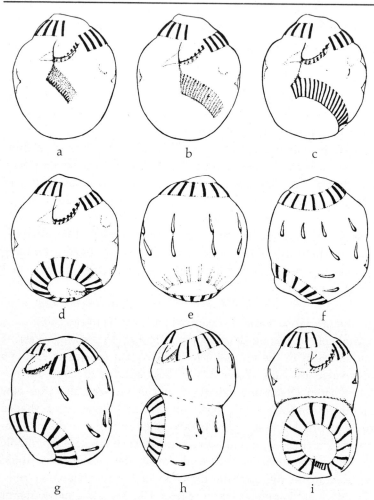

Fig. 33. Cell division in the ciliate *Halteria*. These drawings are from silver-impregnation preparations and show the ciliary bands. Note their movements and shift in polarity during the course of division. [From E. Fauré-Fremiet, *Archives d'Anatomie Microscopique ed de Morphologie Expérimentale*, 42, 214 (1953).]

the surface were a viscous liquid, an idea that is consistent with modern views that the cell membrane is a "fluid mosaic" and that, for instance, the presence of an antibody will induce the formation of a polar cap made up of immunoglobulins on lymphocyctes which is presumed to be by the mass movement of globular proteins to one region.[17] But what directs the movement in the ciliates; what causes it to be so orderly?

Another example comes from the hypotrich *Euplotes*. On the ventral surface the primordia of the new cirri (which are clusters of basal bodies) expand in two groups, one for each daughter cell, and as they expand the old cirri of the parent cell are reabsorbed (Fig. 34). The expansion is between the cirrus bases; the bases first achieve their maximum size and then separate from one another at a uniform rate.[18] In this case the movement could be by an even intercalation of material between the cirrus bases, but we have no clue to the true nature of the process.

A particularly revealing kind of movement of basal bodies on the cell surface appears in suctorians. Some time ago Chatton and Lwoff[19] showed that in the formation of the ciliated larva (which has standard ciliate kineties) there is not a new synthesis of basal bodies, but the basal bodies are present and irregularly placed all over the surface in the adult. When the larva is formed, the randomly placed basal bodies move into rows (Fig. 35). Again there is a fluid property to the cortex; it is as though one were observing the formation of the cortical macrocrystal from the fluid state. Suctorians are unusual in that they do not pass on the organized architecture of the surface directly at each cell division. They have a more elaborate life cycle in which the cortex plays an important role. While they are sessile adults, they show no symmetrical cortical structures, as we have seen; it appears in the larva and disappears when the larva turns into an adult. Therefore it is not universally true that all ciliates pass on their cortical architecture directly; suctorians redevelop it in each generation (but paradoxically in the larva rather than in the adult).

A particularly interesting case has recently been discovered by Grimes[20] in *Oxytricha*. He has shown that during the encystment of this organism, unlike that of many other ciliates, all the

183

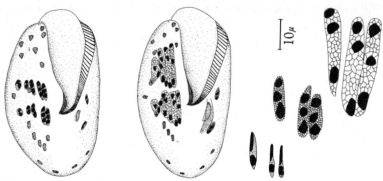

Fig. 34. A silver impregnation of the ventral surface of *Euplotes*. The new groups of cirri are formed by the expansion of two parallel sets of primordia, which appear near the mid region. An enlargement of this expansion is shown on the right. [From J. T. Bonner, *Journal of Morphology*, 95, p. 97 (1954), Fig. 1.]

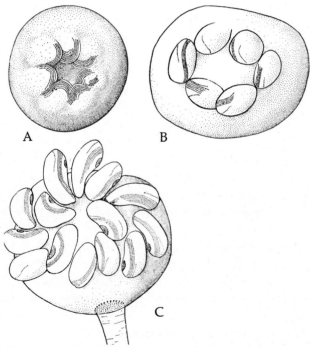

Fig. 35. The formation of motile larvae in the suctorian *Ephelota gemipara* Hertw. The silver-impregnation technique reveals the fact that the basal bodies line up in the formation of the larval buds. [From V. Guilcher, thesis, Université de Paris (Masson, Paris, 1950), 75, Fig. 22.]

184

basal bodies and the cortical pattern disappear. Upon germination there is a total *de novo* synthesis of the basal bodies, and they become aligned in their regular pattern. Even more remarkable is the fact that if doublets encyst they emerge as doublets, even though all adult cortical structures disappear in the cyst.

From both suctorians and *Oxytricha* it would appear that, even though the nucleus could control directly certain aspects of the genesis of the cortical pattern, there are clearly controlling factors that lie in the cortex itself, such as would be inferred from the encysting doublets. The whole process of their forming rows or the disappearance of the rows could be due to changes in the properties of the macromolecules of the surface.

It is important to restate at this juncture that our purpose here is to establish physical means by which substances are positioned within organisms. The first case has been that of specific bonding between molecules; the position is established by the weak interactions between macromolecules. The argument suits the formation of small macromolecular aggregates, and with the help of special proteins probably applies to the positioning of substances within organelles, although already here one is uncertain whether or not this is the whole story. Other factors might also play a part. Simple physical diffusion, for instance, must be important in mitochondria or even perhaps in basal bodies. The argument that the entire cortex of a ciliate is governed solely by weak interactions would exceed the bounds of credibility. This is so simply because ciliate surfaces are very large, and it is inconceivable that other forces do not also play an important role. But how, all together, they manage to expand some areas and contract others, how they move basal bodies or groups of basal bodies from one spot to another, is hard to imagine. Self-assembly is of key importance, but its exact role in the formation of large structures is unfortunately vague.

Cell adhesion

There is an analogous kind of placing of substances that occurs at a higher order of magnitude. It also involves weak interactions, but in this case it is the adhesions of whole cells rather than of macromolecules alone. We still have only rudimentary biochemical ideas of how some cells stick together and others do not, and what controls the strength of adhesion, but it is obviously due to the binding of molecules on one cell surface with those on another.

The idea that cell adhesions might play an important role in the development of multicellular organisms comes originally from Julian Huxley's[21] work, which reinterpreted the early experiments on sponge reassociation of H. V. Wilson. If sponges are pushed through bolting cloth, the isolated cells reassociate rapidly and in a few days one has a new sponge with canals and flagellated chambers. In the early work it was assumed that the cells all reverted to their embryonic condition and those in the inside became collar cells, while those on the outside differentiated into epithelium. Huxley was able to isolate pure collar cells and found that they did not form a new sponge, but merely stuck together as curved sheets of permanently differentiated collar cells. He interpreted this to mean that in Wilson's experiments the cells retained their differentiated character and, by wandering movements, collar cells stuck to collar cells, epithelium to epithelium, and that this "sorting out" was the basis of the reconstitution described by Wilson. As Huxley pointed out, this was the reverse of Driesch's dictum: instead of the fate of a cell being a function of its position, in this instance the position of a cell was a function of its fate, its differentiated state.

The next jump forward came from the work of Holtfreter, and later of Moscona and Weiss,[22] who discovered that scrambled cells from an embryo also kept their differentiation potentialities and sorted out in such a way that usually one tissue surrounded the other. There has been much discussion of the mechanism of this sorting out and at the moment by far the most all-encompassing and satisfactory theory is that of Steinberg.[23] He proposes that the crucial difference between the surfaces of two cell types is simply quantitative; it involves the work, or the strength, of adhesion. If one is mixing two cell types (*A* and *B*) of different cell-surface properties, there is a specific adhesive work between *A* and *A*, *B* and *A*, and *B* and *B* cells. From this one can calculate a variety of equilibrium conditions: for instance, if one assumes that the work of adhesion in AA is greater than in BB, and in AB is equal to or slightly greater than in BB, then one would predict that the A cells will be in a central ball surrounded by B cells. Every way attempted so far by Steinberg and his co-workers to measure these works of adhesion in different tissues has been consistent with the hypothesis. These cell-surface forces are then potentially a method of putting substances in specific places. For example, the collar cells of a dissociated sponge, which differ in their overall chemical constitution from other cells, end up in central aggregates of cells presumably because of their mutual adhesive forces are greater than those of the epithelial cells of the sponge.

There is, however, one added mechanism not mentioned so far that, as Steinberg has stressed, is essential for sorting out. The cells must move, and in fact all the cells that do sort out creep around by amoeboid motion. We shall return later to the role of cell movement in the positioning of substances in developing organisms; here it is added as a "given" in the equation. It may well be that in the macromolecular aggregations, especially those of the ciliate cortex that we just discussed, the element that we fail to understand is some kind of surface locomotory system. Once a macromolecule has moved into the right place by bumping into a complementary macromolecule, it is snapped into position by self-assembly, much as cells that are mutually adhesive will stick together.

One of the difficulties with the sorting-out process is that all the instances in which it has been experimentally demonstrated have been by artificially mixing different cell types; the crucial question is whether or not it is an important factor in positioning substances in normal development. Holtfreter argued strongly that it was, and Steinberg[24] has evidence to suggest that many of the major morphogenetic movements in early amphibian development are governed by these adhesive forces. If this is so, it means that the cells must be different before the normal sorting out; differentiation must have begun before the major morphogenetic movements. There is clear evidence that this is the case in the cellular slime molds.[25] The amoebae before aggregation differ in size and in the amount of spore protein they possess. Immediately after aggregation into cell masses the sorting out occurs, and the cells containing the spore proteins end up in the posterior end of the migrating slug, their correct position for further development. There are two important points about this case. First, these differences are not genetic but somatic; this is a clear case of range variation, possibly because of unequal cell divisions. Second, these early differentiations are reversible, as can be shown by cutting off a fraction of the slug from any part (for example, the posterior, prespore zone) and a diminutive normal fruiting body will appear complete with stalk cells made out of cells that previously were destined to become spores. One must presume, therefore, that the sorting out of the cells serves perhaps to produce a stalk-spore gradient, any part of which can regulate into a perfectly proportioned and differentiated fruiting body. We shall consider this matter in more detail further on.

Another clear-cut and particularly interesting case of naturally occurring sorting out is suggested from the work of L. J. Hale[26] on the hydroid *Clytia*. He has shown that as the colony grows and new colonies sprout from the advancing stolons, the older branching stem with hydranths regresses (Fig. 36). But the cells are not lost; by using markers he showed that they leave the older regressing polyps and migrate through the gastrointestinal canal to colonize the newly forming ones. Therefore new growth consists only partially in cell proliferation and the synthesis of new protoplasm; it also involves the sorting out into new tissue of old cells that arrive separately.

In this instance, and for other animals in general, including vertebrates, it is unclear whether the cells that sort out are first different for somatic or range-variation reasons, or whether they have rather systematically followed a series of steps toward differentiation. The distinction is not profound: in one case one assumes a totally haphazard process such as unequal cell divisions, while in the other case one imagines that certain cells take one avenue of differentiation rather than another because of their environment. For example, perhaps some cells have more yolk in them than others, which leads to certain chemical events that set the cell off on a certain direction of differentiation. The question whether or not these changes are reversible also probably varies with the organism. In some cases they undoubtedly are, while in others it is reasonable to assume that determination occurred so that the cytoplasm is somehow locked in and change is impossible.

Cell adhesion (assuming locomotion) can be an important way

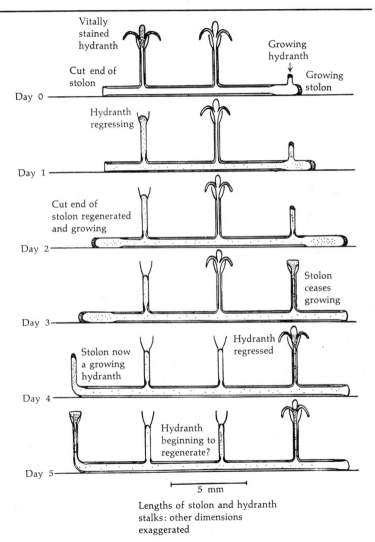

Lengths of stolon and hydranth
stalks: other dimensions
exaggerated

Fig. 36. Movements of cells through the gut cavity from a regressing hydranth to a growing hydranth; dots indicate marked cells observed in the experiment. [From L. J. Hale, *Journal of Embryology and Experimental Morphology*, 21 (1964), text figure 7.]

of localizing substances in development. Since it involves cell-surface adhesions which are affected by the molecules at the surface, it is a kind of supercrystallization. But because of the locomotory powers of the cells, the shapes produced will vary enormously. Another factor in affecting shape is (as in macromolecules) the possibility that there may be sites or regions on the cell surface that show different degrees of adhesion. In rigid cells this is especially obvious; even the differences in the radius of curvature of different parts of the cell could alter the adhesion properties. An excellent example comes from the green alga *Pediastrum*. During the swarming of the zoospores in a vesicle they form flat plates (Fig. 37). If one watches the process closely it is obvious that this is because the cells attach end to end, and the disc is the predicted shape in view of the fact that the elliptical cells tend to adhere at their highly curved ends.[27] There is also evidence for localized zones of adhesion in slime mold amoebae, which, of course, lack rigid cell walls.[28] The question of what role this plays in shaping the cell mass is an unexplored and intriguing one.

Diffusion

The next physical process that we shall consider in the placing of substances in a developing organism is diffusion. This phenom-

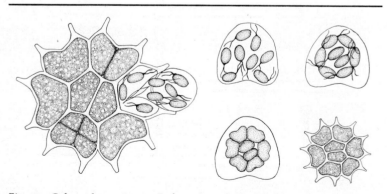

Fig. 37. Colony formation in *Pediastrum:* (*left*) a mature colony undergoes division in some of its cells, which are then liberated in a vesicle (*right*), and the swarmers proceed to form a flat plate that becomes a new colony. [Based on drawings of G. M. Smith and J. G. Moner.]

enon is ever present in all living systems and clearly must play an important role. That sufficient energy can be derived from a diffusion gradient to power cell movement and cell division was first suggested by Rashevsky.[29] This does not mean that this is the source of power for these cell activities, but only that it is theoretically possible.

In order to have a diffusion gradient one must have an asymmetric source of production of a substance. If perfect equilibrium were reached, all substances would be evenly distributed. Therefore, if diffusion is to be an effective factor in the movement of substances, one must assume a basic asymmetry in the cell or cells, that is, an active production of substances in localized regions as a result of the expenditure of chemical energy provided by metabolism. These asymmetries may be produced by a specified region of production of the substance, or by localized permeability of the cells to the substance. These are the basic conditions for Rashevsky's arguments.

Now let us add to this, putting particular emphasis on the possibilities in multicellular organisms. There can be an enzyme that destroys the substance, and the distribution of the enzyme can be uniform or asymmetric, both of which conditions will alter the diffusion gradient in particular ways. The substance can also disappear by natural decay, or it can affect its own production in a feedback loop, either by inhibiting or by stimulating its further production. P. A. Lawrence[30] has suggested that such a gradient might exist in the formation of the cuticle in the bug *Rhodnius*. He proposes, following an argument of Crick, that all the cells in a gradient are capable of producing the substance and that they do so at each point at the level set by the original gradient. This he calls a "homeostatic" gradient. It is also conceivable that a gradient of this sort exists for acrasin in the multicellular stage of the cellular slime mold. Here there is evidence for progressively less acrasin produced by the cells as one moves posteriorly along the slug. As development proceeds, however, the posterior cells slowly lose this homeostatic ability, and as the slug approaches stalk formation, only the anterior cells appear to be producing acrasin.[31]

191

There is another possibility, suggested by B. M. Shaffer,[32] which postulates a totally different kind of diffusion mechanism. In the cellular slime molds a cell can produce a puff of acrasin, creating a temporary small gradient in its immediate surroundings. This gradient spreads, and as it hits a neighboring cell, it stimulates that cell to produce a puff of acrasin in turn. This process repeats itself on the next cell in line, and so forth. If one now assumes that after such a puff each cell undergoes a refractory period, one has a wave of high acrasin concentration that passes along a line of cells. The cell that pulses most rapidly or begins first will be the pacemaker; the whole process in this scheme, as Shaffer points out, closely follows the pattern of a succession of nerve impulses along a nerve.

Crick[33] makes a distinction between these two kinds of diffusion systems by calling the former random-walk mechanisms and the latter signal mechanisms. Goodwin and Cohen[34] have developed an interesting and sophisticated hypothesis for a signal mechanism in which two signals are given off at slightly different frequencies and the various regions of the organism can respond in different ways to this phase difference (one is reminded of the old *moiré* theory of development of Sleggs[35]). There certainly is no theoretical reason why some spacing mechanisms could not operate in this way; as always, one would expect a variety of spacing mechanisms by convergence. But whether such phase-difference mechanisms are in fact operating in some developmental systems must await further critical and painstaking experimental analysis.

On theoretical grounds, one of the oldest concepts in embryology is that thresholds in gradients might play a role in development. If a particular reaction is dependent upon a critical concentration of a morphogen, one can expect that one portion of the organism will develop in one way above the threshold and in another way below. Thus it can be argued that even if the organism is bisected, and each part is capable of regulation, there could easily be mechanisms that would so regulate the absolute concentration of the morphogen that two new gradients could appear. The real difficulty with all these gradient hypotheses is

that there are too many possibilities and too few genuinely critical experiments to prove that gradients do exist and are instru-
mental in governing development. Let me give a few instances in
which it is clearly established that gradients of a substance have
an identifiable morphogenetic effect.

The first example is the development of the egg of the brown
alga *Fucus*. Before fertilization this egg is unusual in that it appears to be absolutely symmetric about a point; there are no axes
of symmetry. Ultimately, when it begins to develop, it produces
a rhizoid at one end and later a cleavage plane at right angles to
the axis of rhizoid formation (Fig. 38). There are numerous external factors that affect the point of appearance of the rhizoid,
such as light (the rhizoid appears on the dark side) and temperature gradients, but here we are concerned with chemical gradients,
which will appear in conditions of uniform illumination (or darkness) and uniform temperature. This was first described in the
last century by Rosenvinge, and was pursued by Whitaker in
the 1930's, but the crucial experiments have been the recent ones
of Jaffe and his associates.[36] Rosenvinge showed that if one took
a group of eggs in uniform conditions the rhizoids tended to grow
toward each other; this is known as a positive "group effect."
Jaffe showed that this depended on having many eggs in the
group. If one used isolated pairs under identical conditions (pH,

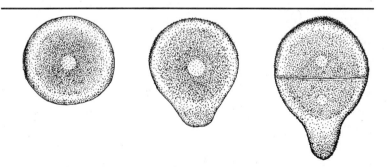

Fig. 38. Rhizoid formation in the egg of *Fucus*. The egg is symmetrical
around a point, with the nucleus in dead center. First the rhizoid appears as
a bump and then the first cleavage plane forms at right angles to the axis of
elongation.

193

and so forth), the effect was reversed, giving a negative group effect. From this Jaffe has proposed a hypothesis that fits all the facts. He suggests two substances: *rhizin*, which stimulates growth and decays rapidly, and *antirhizin*, which is stable and inhibits growth. In a large group there would be relatively little gradient of the stable antirhizin, but at the edge of the group the unstable rhizin would fall off in concentration. Therefore the eggs would germinate with their rhizoids toward the center, toward the high concentration of rhizin. In pairs of cells, both show a negative group effect and point away from each other. In this instance the two substances will have an almost equal chance of producing a gradient; the antirhizin clearly overrules the rhizin and the rhizoids are inhibited between the cells. Jaffe and his co-workers have done a number of other experiments, such as placing individual eggs in low gradients of egg substance, and placing eggs in carefully designed flow chambers so that the key substances can be moved to the downstream side of the egg. They all favor the hypothesis that a gradient of one of the substances is necessary. Furthermore, Jaffe postulates that once the unstable rhizin, which may be indoleacetic acid (and antirhizin may be some enzyme that converts it to an inactive form), starts being produced at the rhizoidal end, it serves as an extracellular rhizin gradient that provides a natural amplification system. The basic notion then is that, beginning with a perfectly spherical egg with a nucleus in dead center, something causes an asymmetry. This may be the fertilization point, light, heat, or its relation to its neighbors or to the vegetative tissue of other algae which seem to produce large quantities of rhizinlike substance, presumably IAA. In some way a difference in one spot on the cell occurs, and once this happens, it becomes a point of production of the growth-stimulating rhizin that soon amplifies itself so that ultimately a proper rhizoid appears at that spot.

Another clear and established case of gradients is in chemotaxis. A beautiful system recently analyzed in some detail by Adler[37] is the attraction of bacteria (particularly *Escherichia coli*) to various organic substances. Again in this case cell movement is a key part of the system and here we shall be considering only the gradient. Adler has provided good evidence that there are receptor sites on the surface of the bacterium for specific attractants. Most of these attractants are metabolites, such as simple sugars, but mutants can be obtained that are incapable of utilizing a particular sugar, yet nevertheless the cells will retain the ability to move up gradients of the substance. He also has worked on the permeases of these sugars, and in mutants that lack the permease the cells are chemotactically more sensitive, suggesting that the important events occur at the cell surface. If the surface is surrounded by a higher concentration of attractant (or an increasing concentration as it swims up the gradient), it will be stimulated to keep swimming in the same direction; if it is surrounded by a decreasing concentration, and fewer of its receptor sites are bombarded, then it will tend to stop, turn, and move in another direction.

In the case of E. *coli* the effect of this chemotaxis is not developmental; it simply distributes the bacterial cells and causes them to accumulate in clumps around attractant substances. Chemotaxis, which is responsible for the movement and placement of substances in development, is well illustrated in the cellular slime molds. Here the amoebae are at first evenly distributed over the surface after they have depleted the food supply, and this even distribution breaks up into a series of aggregation patterns (Fig. 39). Each of these has a center and collects streams of amoebae from the area surrounding that center. In some species it is now known that the attractant (the acrasin) is cyclic AMP and that there is also an extracellular acrasinase which is a specific phosphodiesterase that converts cyclic 3',5'-AMP to 5'-AMP. It is clear that concentration gradients of this acrasin are responsible for guiding the motile amoebae into the central collection points. Presumably they can do this both by overall gradients and by small temporary gradients such as Shaffer suggested and we have described previously. In this way it is possible to orient moving cells and therefore control the movement of substances into a cell mass. Whether chemotaxis plays a further role in the cell mass is not at all clear; we simply have no information.

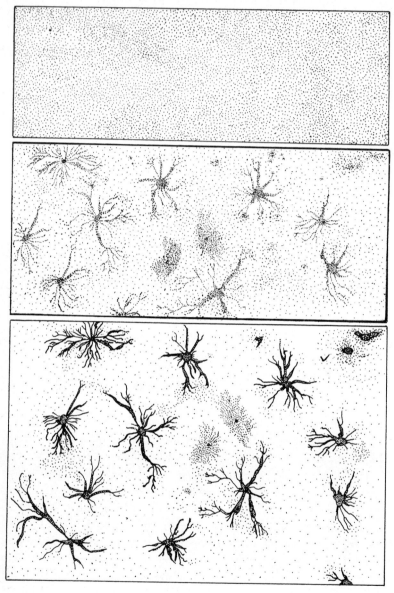

Fig. 39. Three different stages (semidiagramatic) of aggregation of the cellular slime mold *Dictyostelium*.

The entire concept of gradients and their role in development owes most to the work of C. M. Child.[38] First using metabolic poisons and later oxidation-reduction dyes, he showed that in any developing system one region had a particularly high metabolic rate. The importance of this work was the demonstration that these metabolic gradients were universal. Perhaps the best way of illustrating his point is to look at the work of Child and his students (particularly L. G. Barth) on the role of these gradients in the regeneration of the colonial hydroid *Tubularia*.[39]

If a portion of the long stem of *Tubularia* is cut into small pieces of equal length, it can be shown that the more apical (or distal) the piece, the more rapidly it will regenerate a new hydranth. Furthermore, its oxygen consumption can be directly measured and is found to be highest in the apical piece and to fall off for the segments as one progresses posteriorly; there is a perfect correlation between the rate of regeneration and the rate of oxygen consumption. (Again this suggests a homeostatic gradient in the sense of Crick and Lawrence.) If one artificially alters this rate by reducing the oxygen supply, then one reduces the rate of regeneration. This is an old experiment of T. H. Morgan's.[40] He put stems of *Tubularia* into the sand, and new hydranths formed at the end sticking out, regardless of their original polarity. In other words, if the anterior, high-metabolism end is buried, the posterior end now becomes the active end, because it has more oxygen available to it and therefore soon has a relatively higher metabolic rate. By sticking the stems in holes in a plastic partition between two small aquaria Miller[41] was able to show unequivocally that if the oxygen tension of the water was sufficiently high on one side of the partition as compared with the other, he could reverse the polarity and cause the posterior end to regenerate a hydranth. (There is another factor besides oxygen that can also produce this effect, which we shall consider momentarily.)

From this kind of experiment Child proposed his concept of "dominance." The end that was high in metabolism (either natural or induced) dominated the other regions. He presumed that it did so by grabbing the substrates needed for metabolism and in so doing it automatically impoverished the neighboring regions. He pointed out that if the piece of stem of *Tubularia* was short, then only a hydranth was formed at the anterior end; if it was a long piece of stem, a new hydranth appeared at the posterior end also. From this he argued that the stem had become too long for the anterior high-metabolism region to exert its dominance; therefore "physiological isolation" was established and the distant end could set up its own metabolic gradient.

There followed a considerable amount of work on how this dominance effect was mechanically achieved in the organism. The stem consists of a cylindrical tube of tissue (ectoderm, mesoglea, and endoderm) and the center of the tube is the hollow, gastrovascular canal. Is the competition for substrates through the tissue or through the central canal? Barth showed that an oil droplet in the lumen of the canal blocked the dominance to an appreciable degree, while if he inserted a minute glass tube in the lumen and ligated the tissues around the tube so that the gastrovascular canal was the only link, the dominance was not blocked. He concluded that there must be substrates in the canal that were preferentially grabbed by the high-metabolism end. Since then Miller[42] has done an experiment that has put this interpretation in doubt. He built a clever device in which he was able to suck seawater continuously through a regenerating stem and he obtained 100-percent regeneration. There were no substrates at all, yet the original high-metabolism ends of each stem regenerated in a normal fashion. Clearly, then, competition for substances in the canal is not the sole explanation; there is in fact a wholly different explanation of dominance involving inhibitors. But it must be understood that dominance is not explained by either oxygen gradients or inhibitors alone, but by both. The organism seems to be taking no chances and has two systems of achieving the same thing.

The first evidence of an inhibitor comes from a reinterpretation of Morgan's early experiments. Not only did the sand occlude the oxygen supply, but also it could have prevented an inhibitor from escaping. There has been some controversy in this approach, but it is clear that there is an inhibitor, and Rose and Rose[43] showed that water in which regenerating stems of *Tubularia* had been sitting does inhibit the regeneration of hydranths on new, fresh-cut stems. Numerous other authors have also found evidence for inhibitors in hydroids and there is now a considerable body of evidence to support the notion that one region can maintain its dominance over another by a specific inhibitor. It is an exact parallel (as we shall stress later) to the kind of inhibition of secondary reproductives (and other castes) found in social insects.

Another excellent example of the same phenomenon is to be found in the apices of higher plants; dominance can be shown to occur at two levels. One is on the microlevel, where the growing points of leaves (or sepals and petals) achieve a spacing, a symmetry known as phyllotaxis. From the earlier work of Wardlaw and others using surgical techniques there is good evidence that the position of new growing points is determined by the exerting of a dominance effect by the older growing points. This conclusion is supported by the recent experiments of Schwabe,[44] who treated plants having spirally oriented leaves with tri-iodobenzoic acid (TIBA), a substance that affects auxin activity and transport, and the plant apices began producing alternate leaves (Fig. 40, *above*). Since the TIBA traetment greatly increases the length of the apex, Schwabe suggests that the change in phyllotaxis is caused simply by the vertical distribution of the diffusion zones; in the TIBA plants the zones become too far separated from one another to produce the spiral pattern (Fig 40, *below*). The diffusion zones could be (as we indicated before) regions of competition for substrates, or the dominant points could produce an inhibitor that prevents a new growing point from appearing within a certain radius. Again there is no theoretical reason why both mechanisms might not operate.

Dominance at the second level is the inhibition of lateral buds by the apical bud in certain plants. It used to be thought that this was easily interpreted in terms of the distribution of auxin, but now the situation is known to be more complex.[45] There are

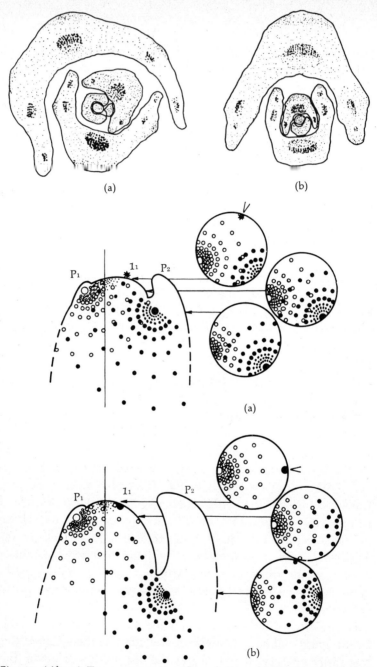

Fig. 40. (*Above*) Transverse section through the apical bud at the level of the bare apical dome, (*a*) control, (*b*) after TIBA treatment; (*below*) possible diffusion patterns for inhibitors as seen in longitudinal section, and transverse sections at three levels (diagrammatic). [From W. W. Schwabe, *Society for Experimental Biology, Symposium* (Cambridge University Press, Cambridge, England, 1971), vol. 25, Figs. 1 and 8.]

clearly a number of hormones involved, probably both kinetin and giberellin, and some inhibitors of an unknown nature. The distances over which these inhibitors can act are much greater than in the case of the determination of the pattern of phyllotaxis, but the basic principle appears to be similar. It is not clear how important it is, in these various cases, to have the key substances in gradients; one presumes that this might be so in the control of phyllotaxis, but the case is less compelling for apical dominance.

It is impossible to leave the subject of gradients and development without mention of the double-gradient system of sea-urchin early development.[46] There are clearly two different kinds of metabolism: one is characteristic of the animal pole of the egg and the other of the vegetal pole. As was mentioned earlier, Driesch showed that if a sea-urchin embryo (or the egg) is cut through the animal-vegetal axis, two perfectly proportioned dwarf embryos are formed. If the cut is made transversely, the vegetal half forms a simple gastrula and stops, whereas the animal pole forms no more than a ciliated blastula. For molecular reasons that are not understood, some agents, such as lithium chloride, will inhibit the animal pole and convert a normal egg into something like a vegetal half; and other agents, such as sodium thiocyanate, produce an animalized embryo. In the beautiful experiments of Hörstadius, he recombined different blastomeres in an early embryo and again found that normal development could be achieved only if there was a reasonable balance of cells from the two poles. The importance of this system is that it shows clearly not only that there is a possibility of a gradient in embryos, but that there can be two opposing gradients that are both necessary and that exist in an equilibrium state.

In this discussion of diffusion mechanisms we have made the following additions to the notion of a simple diffusion gradient. We have in many instances assumed a continuous, localized source, and also suggested that in some cases there may be a sink, such as instability in the morphogen or its inactivation by an enzyme. We have also suggested that sometimes a gradient might be maintained by a feedback mechanism all along its

201

course which keeps the level of morphogen constant at each point. Finally, we have shown that diffusion of a morphogen can act as a signal and, with the help of refractory periods, move as an impulse down a chain of cells. Now we shall consider some other methods of substance movement that do not involve diffusion.

Polar and active transport

It has been known for a long time that auxin (and more recently other hormones) move in the shoots of higher plants far more rapidly than one could account for by diffusion.[47] Furthermore, the movement is polar, that is, it is mostly (although not entirely) in one direction only. For instance, auxin moves in the stem of a plant from the apex toward the base at a rate of about 10 cm/hr. Recent work of Jacobs and his collaborators[48] has shown the remarkable fact that this movement can occur in dead plant tissue at the same rate, but that auxin moves equally readily in both directions. The conclusion is that the rapid movement is achieved by some physical phenomenon other than diffusion, such as the spreading along an interface between cells. Polarity is achieved by blocking this movement in one direction. How the living plant manages this is not clear, to say the least, but it is something that can be accomplished only by actively metabolizing cells.

Numerous organisms have active transport across membranes. There is some evidence for a special receptor protein at the cell membrane which might grab a specific molecule and somehow release it on the other side of the membrane.[49] This is a process that requires energy, and it can, of course, pump against a concentration gradient. Since cell (for example, nerve) impulses involve a loss of ions that have been stored in high concentrations (as well as the reverse), the repolarization of the nerve involves the pumping back of the ions into their original concentration distribution by active transport. There is no reason to suppose that the same mechanism is not used to bring substances through a tissue, and if such a mechanism plays a role in development, this is another way of moving substances that does not involve diffusion.

Recently there has been great interest in the different kinds of junctions between cells in a multicellular embryo. It is increasingly clear that there are a number of different degrees of intimacy of cell contact, leading ultimately to clear anastamoses between certain cells. By electrical studies Loewenstein[50] has shown that many cells in embryos are physiologically continuous and do not have large differences in electrical resistance if electrodes are placed in adjoining cells. This means that there are all sorts of ways in which molecules of different sizes can get from one cell to another. This greatly enlarges the possibility of producing gradients and shifting substances of morphogenetic significance in tissues.

Lawrence[51] has suggested that active transport might, in some instances, be working against a diffusion gradient, the combination producing a particular pattern. One could even imagine the two varying in time: for instance, there could be a spasm of polar transport of a key substance, followed by a return to an even concentration through straightforward diffusion. In such a model the peak of a morphogen would slowly move along the axis of the organism.[52]

Electrical energy

In *Fucus* we find one of the most interesting and important new demonstrations of a way in which key morphogenetic substances could possibly be moved. This comes from Jaffe's meticulous study of the generation of electric currents in the developing *Fucus* egg. There is a long history of the demonstration of electrical potentials in developing systems but they have not generated much excitement over the years because it has been realized that these potential changes simply reflect chemical changes. The matter was in fact put into considerable disrepute because some workers in the past have argued that all of development is controlled by a series of mysterious electrical events.

Jaffe set up several hundred *Fucus* eggs in a capillary tube. These were illuminated from one end so that the rhizoids appeared at the other end (Fig. 41). From this line of polarized eggs it was possible to measure the current generated by each egg, for the total current generated along the tube will be the sum of those of each egg. The question Jaffe then raises is whether or not the currents are solely a consequence of the asymmetric localization of substances by some other cause, or are also directly causing localization themselves. He presents compelling evidence that the currents could be causative as well: the electric current across the cell is sufficiently high to produce an electrophoresis of proteins. Therefore these currents could conceivably be a major force (which Jaffe calls "self-electrophoresis") in moving substances to appropriate places.

Cell movement

We have thus far been emphasizing the physical forces that are directly responsible for the localization of substances, but we have repeatedly suggested that some kind of cell or cytoplasmic movement is often required. Now we shall concentrate on a consideration of the role of cell movement in placing substances in specific locations in development. But first a few general comments about locomotion are in order.

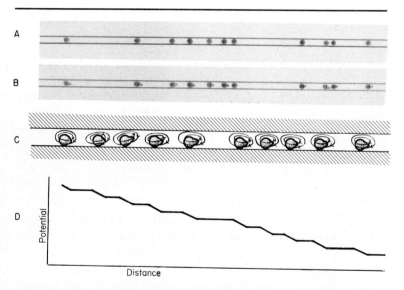

Fig. 41. Jaffe's experiment to show the generation of current in growing *Fucus* eggs: (A) Photograph of eggs, 75 μ in diameter, in part of a capillary with 100-μ bore, before germination; (B) same eggs 26 hours after fertilization; (C) highly schematic view of inferred current pattern in a tube; (D) schematic graph of inferred change of potential along the tube. [From L. F. Jaffe, *Proceedings of the National Academy of Sciences*, 56, 1103 (1966), Fig. 1.]

The fact that cells have locomotory systems is perhaps one of their most remarkable features. They can directly convert chemical energy into mechanical energy with great efficiency. There are deep-seated reasons why this has occurred early in evolution, for the fact that cells have developed a means of locomotion is related to their size. If one plots the size of atoms and molecules against their rate of movement, one can show that as the size of the molecule increases, its spontaneous locomotion (thermal agitation) decreases. This is a basic fact which, for instance, is part of Einstein's law of Brownian motion. One can easily show that once one reaches the size of large macromolecules, they are no longer capable of any spontaneous motion; they move about only because they are bombarded randomly by smaller molecules, which, of course, is Brownian movement. It is safe to say, therefore, that unless living organisms, which begin at this size limit of atomic motion, devised their own means of movement, they would be forever stationary. Many bacterial cells are too large to be affected by Brownian movement.

Those bacteria that do not move, nevertheless survive successfully. It is not exactly true that they do not move at all because growth provides some motion. Cell division involves movement both within the cells and of the whole cell itself, although admittedly the latter movement is modest. Other bacteria are motile and it is significant that there are at least two different kinds of motility, which appear to be totally unrelated. There are those species that move by means of flagella and there are those that have no flagella, but glide.

In eukaryotes one again has two quite dissimilar kinds of cell motion: flagellar and amoeboid. (There are also some gliding forms, such as diatoms.) There are many cells and organisms that are essentially nonmotile, although again these cells move by cell division, cyclosis, and less obvious internal activities; there are all sorts of movements within eukaryotic cells.

The remarkable fact is that, with the possible exception of muscle movement, we still have a very rudimentary idea of how these locomotory systems work. There is a consensus that contractile protein, so clearly involved in muscle contraction, is also involved in amoeboid movement and some of the various kinds of internal movement. Recently there is an active interest in the fact that microfibrils and microtubules are so often associated with movement: they are in muscle, in the mitotic spindle, in cilia and flagella, and in pseudopods of amoebae. There is evidence that these structures are directly involved in the cell movement or cell deformation. It is possible to use agents that will destroy these tubular and fibrillar structures; when this occurs, all motion ceases, and soft cells bounded by a thin membrane revert to a spherical form. No doubt the complete elucidation of how chemically and mechanically each one of these motions is achieved will be forthcoming; in the meantime we have to be content with the fact that at least superficially it is obvious that there are a number of ways in which chemical energy can be converted to mechanical energy in the form of external or internal movement. It may well be that they all have certain basic physicochemical processes in common, although this need not be so. The results are clearly varied: flagellum, pseudopod, gliding, cyclosis, cytokinesis, mitosis, and numerous others.

For the positioning of substances one needs to control this movement so that it is consistent in each generation. We have already discussed some possible evidence for controlled movements within single cells, especially in the cortex of ciliates. But despite the importance of this kind of movement, we have so little understanding of it that for the time being any further discussion is not likely to be helpful. Instead we shall concentrate here on ways in which movement of cell masses is guided. To a large extent we have already done this in our series of examples of chemotaxis and cell adhesion. There are two ways in which mass cell movements can be organized. In chemotaxis one needs to have a chemical gradient which orients the motile cells in the gradient. In cell adhesion the cells move at random, but because of differential adhesive forces between cell types, a sorting out occurs. This can be further influenced by having specific sites on the cell that are adhesive, producing different configurations. It should be added here that there is no reason to rule out the possibility that in some developing organisms (the

cellular slime molds, for example) both chemotaxis and cell adhesion operate simultaneously.

The role of motility in multicellular development can be illuminated by examining the effects of gene mutations (or multi-
gene differences) that affect cell movement. Some years ago
Twitty and Niu showed that the difference in the pigmentation
of two species of salamander was the result of differences in
motility.[53] The chromatophores migrate out from the region of
the lateral line in *Triturus rivularis* and become evenly spaced
(for they have a mutual repulsion or are negatively chemotactic
to each other). In *T. torosus*, on the other hand, the chromatophores never move away from the lateral-line region, giving a black
pigment streak down each side. If these cells are grown in culture,
those of *T. rivularis* spread evenly in the culture dish, while
those of *T. torosus* remain in knotted aggregates. An interesting
parallel was discovered in the axolotl by Dalton.[54] Here the
normal pigmented species behaves like *T. rivularis* and the pigment cells spread evenly, while in the albino form there is no
migration at all. By transplant experiments it can be shown that,
in contrast to the case of *Triturus*, the difference in adhesion and
motility is not due to the cells themselves, but to the tissues
through which the cells migrate. Albino axolotl epidermis does
not permit chromatophores to migrate through it. In both cases
we presume that the gene differences have resulted in chemical
changes either on the chromatophores themselves or on the epidermis which in turn affect their mobility and hence the pattern
of pigmentation.

A more recent case studied by Ede and his collaborators[55]
provides a slightly different kind of example. A mutant in the
chick called *talpid* has the peculiarity of producing limbs which,
at early stages, have the rudiments of numerous digits resembling the fin of an ancestral crossopterygian rather than a more
advanced tetrapod limb (Fig. 42). By removing *talpid* and normal
limb tissues at early stages of development it can be shown that
the mutant cells are more adhesive and less motile. It is presumed
that in some way these cellular differences account for the fact
that in the *talpid* limbs there are more centers of cartilage forma-

tion than in the wild type. These centers have been compared to aggregation centers in cellular slime molds, where numerous external factors (including the acrasin concentration) are known to affect the density of aggregates in a given area.

Growth movement

There is quite another way in which movement can place substances in a consistent pattern in a multicellular organism. This is positioning by growth movements, a phenomenon that is especially important in plants, which have relatively immobile hard cell walls. What is controlled is the direction of the cell divisions (that is, the spindle axis) and the location of which cell divides at a particular time. By varying these parameters it is possible to produce a large variety of patterns. Since I have discussed this in detail elsewhere, I shall here give only the briefest summary.[56] It is possible to get squares (or rectangles) by alternating divisions at right angles to one another in a plane (Fig. 43). If the

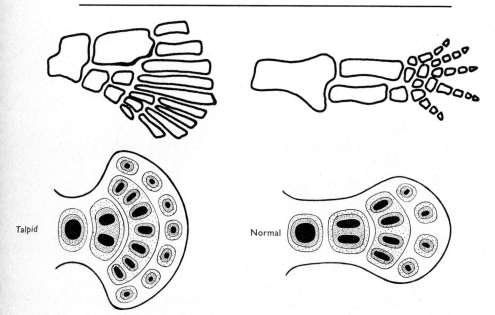

Talpid

Normal

Fig. 42. (*Above, left*) Fin of the Devonian crossopterygian *Sauripterus;* (*right*) primitive tetrapod limb. [After Romer.] (*Below*) Diagram of precartilage condensation fields in (*left*) talpid, (*right*) normal, limb buds. (From D. A. Ede, *Society for Experimental Biology, Symposium.* Cambridge University Press. Figs. 8 and 9, 1971).

divisions, or spindle axes, alternate at the x, y, and z axes, cubes will result. If cells keep the same direction of their spindle axes and are held together by jelly or by adhesive walls, they will form a filament. If occasionally one of the cells of the filament has a division at right angles, then a branching pattern occurs. (There has been some interesting theoretical speculation on the amount of information required to obtain particular branching patterns; in general, a relatively simple set of conditions is sufficient.)[57] If the branches tend to fill the spaces between them, and the organism is growing on a flat surface, a disc shape will result. If the filament tends to have right-angle spindles formed uniformly up and down, a thickened filament will be produced, and if these tangential cell divisions increase, ultimately a hollow tube will result. The alga *Ulva* (sea lettuce) has such a large hollow tube that the two walls have collapsed into a two-layered, broad frond in which cell divisions occur all over the surface in both layers (Fig. 43).

In this evolutionary progression toward large algae one sees the transformation of the filament into a parenchyma type of tissue. This is not the only way in which size increase has been achieved in plants: for instance, in higher fungi the entire mushroom is made up of a mass of filaments; the tissue is a fabric of interweaving threads. Also in some algae, the whole organism consists of one large multinucleate tube, or coenobium. These interesting cases will be considered in detail in the examples given in the last chapter.

In plants with parenchymatous tissue, as the size of the plant increases, there is a corresponding increase in the rigidity of the cells for support, making it progressively more difficult for any cell to divide. As a result, growth zones or meristems have evolved which become controlled areas of the positioning of substances by growth movements. In some algae, such as *Porphyra*, which is a single sheet of cells, the entire outside edge is the growth zone. In various red algae there is an interesting situation in which the complete cell lineages are visible because those cells that have divided leave a telltale thread between them,

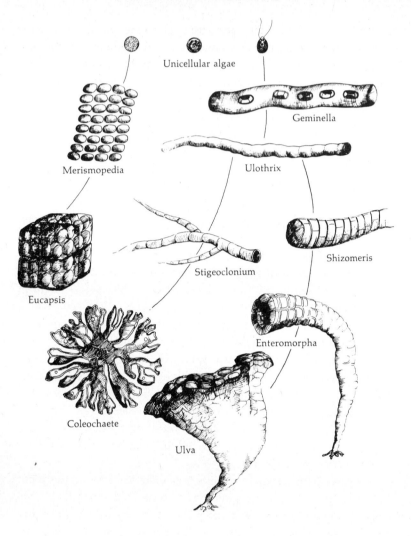

Fig. 43. The various shapes achieved among the algae can be interpreted in terms of growth direction (or cleavage patterns) in organisms with rigid cell walls. By changes in the direction, the frequency, and the distribution of cell elongation and division, a wide variety of shapes is possible. [From J. T. Bonner, *Size and Cycle* (Princeton University Press, Princeton, N.J., 1965), Fig. 4.]

and the sequence of divisions is immediately obvious (Fig. 44). In higher plants the apex is meristematic and it can have either a generalized division zone, or a very orderly one, as in *Equisetum*, which has an apical cell that divides with great regularity. There are also various different forms of lateral meristems or cambia that increase the thickness by adding vascular tissue.

Again in all these instances there arise the questions of what sets the pattern, the distribution of the meristems, or the sucession of divisions of specific orientations. Whatever the answers to these questions, it is clear that the plan for the pattern is a property not of single cells, but of the entire cell mass. In the case of filaments that branch occasionally, it is hard to conceive that any one cell will go through a specified number of divisions

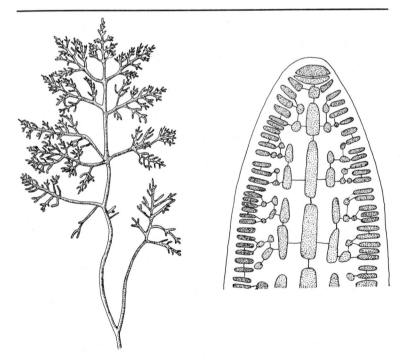

Fig. 44. A red alga, *Gelidium cartilagineum* Gaill.: *(left)* the entire plant; *(right)* diagramatic view of the growing apex of the plant. [Redrawn from G. M. Smith, *Cryptogamic Botany*, Volume I, 2nd. ed. (McGraw-Hill, New York, 1955), Fig. 181.]

in one direction and then produce one at right angles. The branches do not appear in this rigid fashion, but more often seem to be spaced by some physiologically active process (akin to apical dominance). Evidence for this comes from interesting work done on filamentous blue-green algae in which the heterocysts are clearly spaced by substances they give off and thereby influence where another heterocyst will be formed.[58] In general, in all multicellular forms the whole plant has effective intercommunication of parts. Apical growth zones influence where other such zones are likely to appear, as we showed in our previous discussion of apical dominance in plants. Hormones and combinations of hormones will inhibit or stimulate. The pattern they produce is bound to be greatly influenced by the physical and chemical nature of the control molecules and by their means of transport, whether diffusion, polar movement, or some other means. We shall return in the last chapter to this matter of how one sets up the initial states that lead to the ultimate positioning of substances.

There are, however, some cases in which the instructions for the growth movement more probably come from inside the cell than by intercellular communication. This is likely the case for those bacteria or primitive algae that produce squares or cubes (Fig. 43); there is a rigid sequence of divisions at successive right angles (in either two or three dimensions) that one assumes is somehow internally controlled. An even better example is that of unequal divisions as in the formation of polar bodies in animal eggs, or the formation of guard cells that surround the stomata on the leaves of higher plants (Fig. 45).[59]

There is also the bacterium *Caulobacter* which does the same thing: each time it divides one of the daughter cells is flagellated and motile, while the other is stalked and sessile. Recently S. T. Degnen and A. Newton[60] have shown that the stalked daughter cell lacks a G_1 period in its cell cycle; on the other hand, the motile swarmer daughter cell has a G_1 period of some 60 minutes during which the cell loses its motility and develops a stalk (Fig. 46). By using the inhibitor rifampin, which appears to block RNA polymerase, Newton[61] has obtained good evidence that transcrip-

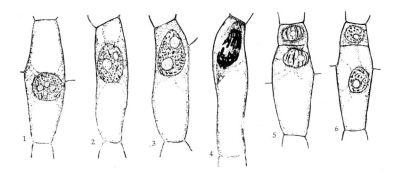

Fig. 45. Stages in the first, unequal division of a cell in anticipation of the formation of guard cells on the epidermis of an onion. [From E. Bünning and F. Biegert, *Zeitschrift für Botanik*, 41 (1953), Figs. 1–6.]

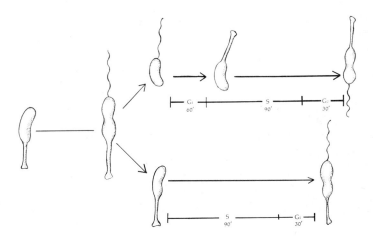

Fig. 46. The developmental pathways of *Caulobacter*, showing the different periods of the cell cycle and the durations of these periods. [After S. T. Degnen and A. Newton, *Journal of Molecular Biology*, 64, 671–680 (1972).]

tion probably occurs for the events that take place during both the G_2 period (for example, cell division) and the G_1 period (for example, loss of motility, stalk formation). Since the latter occur only in the swarmer daughter cell and not in the stalked half, one could assume that the two ends of the parent cell contain different developmental programs. The stalked end consistently produces cell division and swarmer formation, while the swarmer end keeps giving stalks. The key question is how the genomes of the two ends of the cells can differ in their developmental programs: is there some polar structure to the cytoplasm that causes this difference?

There is another area of controlled growth movement that has held a special fascination for many years. It has to do with the growth of nerves in such a way that ultimately the complex network of an animal's nervous system is properly hooked up. There are probably many factors that come into play: some of them we readily understand, but others have resisted elucidation. One factor that clearly can play a part is cell adhesion, as is seen from the work of Sidman and his group on the development of the cerebellum and other parts of the brain in mice.[62] Normally one group of cells migrates past another at an early stage in development and the cells touch in passing. At the points where they touch, dendrites connecting the two cells form and are pulled into long processes as the cells separate. This is the way the interconnections between the cells are established. In some mutant mice that have motor difficulties, the cells fail to move and the dendrite pattern never forms properly.

It is difficult to know in this case to what extent the elongation is growth or a passive pulling of processes from the two cells, as one moves past the other, rather like pulling taffy. In any event, there are many well-established cases in which nerves will send out axons that connect end organs with the central nervous system: for example, motor nerves to muscle, or sensory nerves from the ear or the eye. We shall examine the question of possible mechanisms of the specificity of these nerve connections in the last chapter; here I merely want to point out that there is an organized pattern to the nerves which is set down by controlled growth movements.

Movement in multicellular organisms

The subject of the nervous system leads us directly to the nervous control of movement in multicellular animals. In single cells we talked of radiant energy and chemical gradients orienting cells, but, at least in the latter case, the actual mechanism was obscure. In animals (and again we shall specifically concern ourselves with social insects), there are eyes, chemical receptors, and even ears, which, through the astounding powers of the nervous system, make orientation and communication between individuals comparatively easy for the insect and easy for us to understand, at least in principle.

In order for the parallel to be exact with a developing multicellular organism, we have to clarify what we mean by form in insect societies. One of the most obvious aspects of the form is the shape of nests. In all different groups of social insects there are examples of exceedingly elaborate nests: consider the tiered nests of some of the paper wasps, the complicated tunneling in an ant mound, or the structure of a large termitarium. These forms are analogous to the shell of a mollusc or a foraminiferan; they are not themselves made of living material, but are laid down or manufactured by living organisms. Unfortunately, we do not know very much about the behavioral mechanisms that lead to the formation of nests of a particular architecture. The aspect that has received most attention is the structure of the honeybee's cell, a matter delightfully pursued in D'Arcy Thompson's *On Growth and Form*. But even in this case there are many unanswered questions: we know it is not because bees are advanced mathematicians and have cleverly calculated the most efficient angles, as some have suggested; we know it is probably not the result of the fact that the bee's eye has hexagonal ommatidia, as one early author suggested. (The idea was that the bee saw the whole world in hexagons, so naturally it would pummel the wax into that form.) As D'Arcy Thompson points out, it is more likely that it has to do with making cylinders with rounded ends, and by compacting these together in two layers one gets hexagons with rhombic ends. But obviously one still wants to know more.

Another way an insect colony can have form is in the changing

215

mass aggregations of individuals. This is seen especially well in the army ant, *Eciton hamatum*, which has been examined in such detail by Schneirla and his collaborators.[63] The colony, which is made up of many thousands of individuals, can exist in two forms: (1) a cluster, which is the resting bivouac in which the ants literally hang together in a ball, and (2) a marching column, which consists of many individuals following one another. But at the tip of one of these columns there will be, especially in the presence of possible prey, a fan-shaped advancing edge. This appears to be an effective weapon for producing flanking movements in trapping escaping insects. The worker and soldier army ants are essentially blind, their eyes being rudimentary. They rely heavily on chemical communication in creating these forms. They lay down trail substances, and other ants follow these trails slavishly. The mechanisms of trail making in other ants have been studied in some detail in Wilson's laboratory, and many properties of the substances, including the factors affecting their "fadeout," that is, their disappearance, are now well understood.[64] In army ants, Schneirla showed that if the ants were placed on a path of bare ground, they would lay trails in circles around the mass of fellow workers, then go around and around following one another until they perished. The fan-shaped front of the column is thought to be largely due to the reluctance of any ant to take the lead. Those in front are literally being pushed forward by the followers from behind. As a result the advancing edge broadens. They can, of course, sense food, and will be attracted toward it, overcoming the reluctance to go where no ant has already been. Less is known about the signals that herald the formation of the bivouac cluster, but clearly dwindling light must trigger the event (since it occurs each night) and no doubt chemical signals are also involved.

These forms, when compared with those of single-celled or multicellular organisms, are relatively amorphous. They involve simple signals in which special chemical substances play a particularly important role. Even though the properties of the chemicals are of prime importance, one also has to be concerned with behavior mechanisms that respond in specific ways to the chemicals, as well as the behavioral responses that result in the precise depositing of the substances. In other social insects, and especially in vertebrate societies, other signals are equally important: radiation (vision), sound waves (hearing), and pressure (touch). In all three cases the signal is simple, but because the nervous systems of the individual organisms in the society are so complex, an enormous variety of information can be imparted. The principles of form making are essentially the same as in the lower hierarchical levels, but the permutations are increased, and at the same time the form itself becomes less rigid. In these animal societies the individuals are not attached, but free moving. This difference is again due to the effectiveness of their receptor system; they can pass signals across great distances, while the cells of a developing multicellular organism have to keep in close contact. Social animals can even signal quite elaborate information: for instance, von Frisch's bees can tell the other foragers in the hive not only that they have found honey, but where it is; a bee can give the direction and the distance of its find. As the ways of passing information improve (which is simply a more sophisticated way of positioning or localizing substances), the form of the overall unit becomes less rigid. The activities of the individuals become the form, just as the activities of electrons give an atom its form.

Summary

Substances can be localized in developing organisms by molecular bonding, cell adhesion, diffusion, polar and active transport, electrophoresis, cell movements, and growth movements. In the development of animal societies there is the movement of individual organisms which is controlled by a complex nervous system that communicates by auditory, visual, and chemical signals. Now comes the question of how these localization mechanisms, along with the temporal sequence of syntheses, produces a complete patterned individual (or society).

9

The Control of Pattern

We have now finally come to the point where we can put things back together again. In this second half of the book we have been concerned with the mechanics of development, and so far we have isolated many of the elements that go into these mechanics. We have discussed the mechanisms whereby substances are produced, when they are produced, and how they are localized. With these elements we must now return to the whole organism to see how the entire life cycle is so consistently and so perfectly patterned. We are therefore looking at the grand questions of development, the ones that have stirred the imagination of so many embryologists for so long, and the ones that to this day we find confusing and frustrating. The complaint is either that we have made no progress in the last hundred years (or since Aristotle!), or that anyone who believes he sees the light is either a vitalist or someone who is deluding himself with some pet theory that either explains only a small portion of the problem or is so hypothetical that it is of little concrete value.

During the course of preparation of this book I have spent some time rereading earlier treatises on developmental biology. It suddenly struck me that there is an element in the human nature of developmental biologists (myself included) that I had never realized before, and now that I see it I find myself mildly scandalized. We apparently like mystery; we like to think that what we are doing is just a bit more complicated than the tasks of others, and we seem to take some pride in this; we gloat over the fact that some aspects of our problems have so far defied explanation. We revel in the complexity of it all: how is it possible that a bud can turn into a flower, or a fertilized egg into a complex human being?[1] Hans Driesch was quite honest about it: he simply said it is too marvelous to have anything but a vitalistic explanation. Although most of us scorn this view, there is considerable satisfaction derived from the fact that development is an involved process, and the simple ideas of the molecular biologist appear (if one has decided so beforehand) to fall miles short of what is needed for a proper understanding. I could give many examples of this interesting phenomenon, but it would be a digression, for after all this is a book on development and not on the psychology of developmental biologists.

As a counterirritant I shall go to the other extreme and take the position that all developmental processes have rather straightforward explanations. It is quite true that in many, perhaps in the majority of instances we do not yet have enough facts to know what those explanations are, but we can bridge the gaps by making hypotheses or models. In fact, we have arrived at the stage where models are useful to suggest experiments, and the facts of the experiments in turn lead to new and improved models that suggest new experiments. By this rocking back and forth between the reality of experimental facts and the dream world of hypotheses, we can slowly move toward a satisfactory solution of the major problems of developmental biology. So our watchword in this last chapter, which considers the grand themes of development, is never to admit mystery, defeat, or chaotic complexity, but with Calvinistic zeal put such easy, backward thoughts to one side and bravely make a hypothesis at each breach. There is nothing evil about a hypothesis. It may be foolish or wrong; if it is, then it should stimulate some perceptive person to make a better one. But good or bad it pushes one forward in one's experimental analysis, and facts are like infantry; they alone can win the battle.

Another burden we carry is the feeling that all development

219

must be contained in a unitary theory. In some ways this accounts for the enormous enthusiasm for a genetical approach to development. Genetics is a model of a science with a unitary theory. This began with Mendel and blossomed with dazzling splendor when, in recent years, the structure of DNA and its role in protein synthesis were finally understood; not only was there a set of simple rules to govern heredity, but now we know their molecular basis as well. We have accepted the view that DNA and its activities are also the basis of developmental biology. But this is in no way a unitary theory such as we have for heredity. The reason is that the "activities" are extraordinarily diverse.

A general statement

We are now faced with the task of organizing into a general statement the enormous mass of information we have on how life-cycle development produces a consistently patterned and proportioned individual. This is difficult because, although in the development of different species they might have similar adaptive functions, they may achieve this in radically different ways. Already many examples have been given, but there are innumerable others; the condition permeates all aspects of development. It is obvious even in the broadest sense: consider the fact that the localization of substances in higher plants occurs primarily by growth movements and the polar transport of hormones, while in animals it occurs by morphogenetic movements, including sorting out, and the transport of the hormones is for the most part nonpolar. Nothing could be more different in the way of mechanisms, yet the end result, in terms of repeated life cycles, or even in some instances in form (as in colonial hydroids and plants), is often very much the same.

We begin our statement by a reminder that each life cycle starts with two kinds of information: that which can be taken directly from the DNA, and that which was previously derived from the DNA and lies in the various cytoplasmic substances and structures of the cell. This information is capable, in each cycle, of consistently producing the right kind and amount of substance (Chapter 6) at the right time (Chapter 7), and in the right place. This last aspect was begun in the previous chapter, where we discussed the mechanisms of localization; now we shall see how a proportional pattern is achieved.

There are two very broad aspects of the mechanics of pattern formation: (A) how the pattern starts in the first place, and (B), once started, how it is completed, that is, reaches equilibrium.

A. Pattern Initiation

Pattern arises in three ways: (1) it can be induced or imposed on the organism by the immediate environment; (2) it can arise spontaneously; and (3) it can be directly inherited from one generation to the next.

1. *Externally induced patterns.* Some organisms inherit an ability to respond to the environment during their development. An ideal example is the egg of the brown alga *Fucus*, with its nucleus in dead center; when it is shed, the establishing of an axis of polarity is a major first step in pattern formation. We have seen that this can be done by chemical means with rhizin and antirhizin, and hence the group effect (Chapter 8). It can also be induced by unidirectional light, for rhizoids grow on the dark side. It is particularly interesting that the egg, following fertilization, goes through a period when these external forces can have their effect; there is a special sensitive period, after which the fertilized egg is no longer influenced by any external changes. Furthermore, Jaffe[2] has shown that this sensitive period differs for different external effects; for instance, the sensitivity to chemical gradients occurs immediately after fertilization and falls off rapidly, while the light-sensitive period does not begin until 5 hours after fertilization (Fig. 47).

It is noteworthy that the pigment that absorbs the light is already arranged in a specific pattern. Thus the light effect of positioning a substance is preceded by an orderly arrangement of the receptor system, which presumably occurs by some sort of self-assembly mechanism. Jaffe[3] showed that if the eggs were illuminated from above with polarized light each egg produced two rhizoids 180° apart in the plane of polarization. From this he concluded that the pigment molecules were long molecules, which lay in the cortex of the egg parallel to the surface (Fig. 48). In this way the receptor pigment molecules lying in the equatorial region would receive the most illumination, since they are oriented in the same direction as the plane of polarization; the molecules at the poles would receive the least since they are at right angles to that plane. Since rhizoids form on the dark side, one would predict that they would appear at both poles, under uniform polarized light from above, and this is precisely what happens. There are cases in other organisms in which the evidence indicates that the pigment is oriented at right angles to the surface of the cell, and, of course, there are many instances in

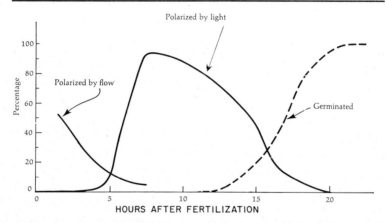

Fig. 47. Time course of the polarizability of *Fucus* eggs by light and flow. [Modified from L. F. Jaffe, *Advances in Morphogenesis*, vol. 7, Fig. 6, 1968).]

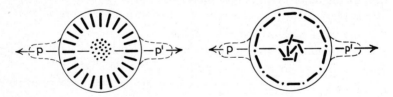

Fig. 48. Models showing the orientation of photoreceptor molecules in the cortex of (*left*) the fungus *Botrytis* and (*right*) *Fucus* (or the spores of the fern *Osmunda*). In both cases they germinate parallel to the plane of vibration of polarized light. This is because *Botrytis* spores germinate on the light side, while *Fucus* eggs and *Osmunda* spores germinate on the dark side. [From L. Jaffee and H. Etzold, *Journal of Cell Biology*, *13*, 22 (1962), Fig. 5.]

222

which the growth point occurs on the illuminated rather than the dark side (Fig. 48).

Once the axis of polarity is established, a series of steps follow that lead to further pattern. As Jaffe has shown, the initial chemical gradient across the egg is a system of self-amplification and this produces an electrical field that soon builds up to such a strength that electrophoresis could occur. The whole further distribution of substances follows from the initial external push of light or of a chemical gradient.

Another excellent example comes from the experiments on polarity reversal in hydroids (also discussed in Chapter 8). Here both the presence of high oxygen and the absence of the inhibitor stimulates either end of a piece of stem of *Tubularia* to regenerate a hydranth. A number of workers have also shown that if one makes a rubble of cells of a hydroid, the cells will reassociate and form a new polyp, but again the mouth and the tentacles will always form on the upper side of the cell mass.[4] The oxygen and the inhibitor gradient impose an asymmetry on the tissue, regardless of its original polarity. There is considerable evidence that one of the first effects of such external gradients in regeneration of a segment of stem is a migration of cells toward the high-oxygen, low-inhibitor region.[5] So again we have a mixture of the localized synthesis of substances and the movement of substance, in this case both induced by an asymmetrical environment.

2. *Spontaneously appearing patterns.* There are some particularly important patterns that appear spontaneously in a life cycle. It is a shade misleading to say "spontaneously," because clearly the basic ingredients for the pattern are there; here we are merely talking about the push that starts the process. The only aspect that is spontaneous is the appearance of the pattern itself, once the initial conditions are given. Let me illustrate my point with a simple example in the nonliving world. If a saturated solution of a sugar or a salt is slowly cooled, there suddenly comes a point where crystals begin to form, and they will produce a beautiful pattern over the bottom of the dish. The given conditions are the saturated solute and the solvent; the change is the cooling; and

the pattern is the crystals and their distribution in the dish. There is a further factor involved here; there must be some perturbation, some unevenness, which may be added quite by chance, that precipitates the pattern. In a supersaturated solution this may be a seed crystal dropped into the solution; even a particle of dust will set the whole process in motion. A uniform equilibrium (the solution) becomes unstable and soon reaches a new, patterned equilibrium (the crystal). This is the sense in which I am using the word "spontaneous."

The mathematician A. M. Turing[6] was the first to systematically apply to biological development the idea that patterns could appear by a homogenous state becoming unstable and shifting spontaneously to an uneven state in which a regular pattern emerged. He developed a model for the formation of a ring of tentacles in a developing hydroid which postulated a key morphogen that went from an even distribution within the cylinder of the hydroid stem to a series of high and low points, which ultimately lead to the formation of tentacle primordia. He also suggested that a similar model might be possible for plant phyllotaxis.

Here I shall describe a biological illustration that appears to fit Turing's concept more obviously than either hydroids or higher plants, and at the same time shows the pertinence of the analogy to crystallization. In the cellular slime molds, after the food supply has been consumed, there is an even distribution of amoebae. Ultimately they will aggregate into a series of aggregation centers, which will be moderately uniform in size and fairly evenly distributed over the surface of the substratum (Fig. 39). This occurs by chemotaxis; as we have already seen, the amoebae secrete acrasin (which is, for some species, cyclic AMP) and an acrasinase (a cyclic AMP phosphodiesterase), and they manage somehow to break up the uniform distribution of acrasin-producing amoebae to form the pattern of centers. Keller and Segel,[7] in a theoretical study, have pointed out that this process can be treated mathematically as beginning with a uniform equilibrium state (the evenly distributed amoebae), which becomes unstable and enters a new state (the pattern of aggregation centers). The changes needed to achieve this instability in the equilibrium are (i) a change in the sensitivity of the amoebae to acrasin, (ii) a change in the rate of production of acrasin by the amoebae, and (iii) a change in the rate of random motility of the cells. There is very good evidence that all three of these parameters do change as aggregation approaches.[8] If the amount of acrasinase is also changed, for example, by inhibition, as Gerisch and his co-workers[9] have suggested, this would be another effective stimulus (for it affects the amount of acrasin produced). It is also necessary to assume some kind of perturbation to initiate the process of aggregation. The usual assumption is that, by random chance, some cells are more advanced than others in their development, and these, because of their precocious secretion of acrasin, become the necessary nucleation centers for initiation.

One of the features of a whole area of aggregates is that the territorial size of each can be constant over a wide range of amoeba densities.[10] This means that if there are many amoebae there will be larger aggregates (therefore larger fruiting bodies), but the distance between aggregation centers will be constant at different densities. According to Keller and Segel, it is possible to explain such constant territory size entirely in the framework of the instability theory, but, as we have already seen, numerous factors will contribute: the removal of cells between centers will decrease the chances of new centers forming there; centers will disintegrate and join a larger center if they lie in a steep acrasin gradient; there is clear evidence for a separate inhibitor so that one center will inhibit the formation of new centers within a certain radius.[11] There are therefore many factors which together ensure some constancy of territory size. One might assume that, once the initial uniformity is shattered by the original perturbation, one has a series of "nucleation" points, points of high acrasin production. These points not only will produce steep peripheral gradients, but could generate a localized increase in the inhibitor. There will perhaps be a period of fierce competition that ultimately will result in fairly uniform territories.

It is important to understand in this example that it involves both localized synthesis and directed movement. The localized synthesis is shown by the excess of acrasin in the initial centers and the production of the inhibitors in the centers. The directed movement is, of course, the positive chemotaxis to the centers, which redistributes the substances over the entire area into the central collection points.

3. *Inherited patterns.* The commonest way in which a pattern appears in early development is by direct inheritance from the previous generation. For instance, most eggs are asymmetrical and when they are shed from the ovary they already show a polarity. (The egg of *Fucus* is unusual in that it is symmetrical around a point.) The obvious ‘cases of asymmetrical eggs are those in which there is a considerable amount of yolk which distinguishes the vegetal from the more protoplasmic animal pole. This ancient point is beautifully seen in the eggs of coelenterates, for when they cleave the nuclear, animal pole furrows first and slowly creases down into the vegetal pole; the yolk distribution determines the direction of the first cleavage plane, and therefore of the subsequent ones (Fig. 49). The classical and obvious explanation is that in the formation of the egg in the ovary, the deposition of yolk was one-sided; the axis of symmetry is directly acquired, or inherited from the mother. Here is one more example of the inheritance of information, other than direct DNA information, that is passed from one life cycle to the next.

A more elaborate example of the same phenomenon (which we

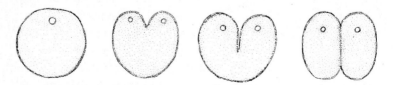

Fig. 49. The first cleavage of a coelenterate egg.

have already discussed) is the transmission of the cortical pattern from one generation to the next in ciliates. In this instance there is normally no great loss of pattern between generations, as there is in an organism arising from an egg or a spore, and hence the case is rather special. In ciliates, not only will the polarity be passed on to the next generation, but so will the intricate pattern of all the superficial organelles.

The initial symmetry or polarity is generally considered as the first sign of development in the life cycle. Here we are merely saying that often the asymmetry comes *in toto* from the previous generation. Polarity is a very general term which simply implies a directionality to the system, and does not imply any special physicochemical basis. In developing eggs there could be a macromolecular orientation or a gradient of some key substance, or both. In fact, one might quickly lead to the other. Many years ago C. M. Child[12] championed the view that quantitative differences would lead to qualitative differences, and at that time he was severely criticized for this view: how could all the beautiful complexities of differentiated pattern be explained in such a simplistic manner? We no longer find any theoretical difficulty with such a concept, and in fact it is a basic premise of modern developmental biology.

B. Pattern realization

We shall consider three general cases: (1) the living units are separate in space (as in the social insects); (2) the living units are attached (as in multicellular organisms); and (3) all the pattern occurs within a single-celled organism. From these considerations certain general principles arise: all patterns are achieved by a system of chemical signals that may be transmitted in a variety of ways; these ways vary with size, and furthermore the greater the distance over which the signals operate, the cruder the pattern.

1. *Pattern in living units that are separate in space.* We shall first examine one aspect of pattern control in social insects: the determination of the proportion of different castes in a colony. This is a developmental problem of special interest, for, among other things, at its simplest, it involves not a fixed pattern in

227

space but a control of the relative numbers of individuals of different castes.

In the primitive case it appears not to be a ratio that is produced but an absolute number. This is the example we have already examined in *Kalotermes*, in which there are two reproductive individuals, a male and a female, and all the other individuals are workers. The ratio will obviously vary with the size of the colony, since the number of reproductives is fixed. As we saw, the reproductives exude an inhibitory pheromone that prevents the other molting nymphs from developing into reproductives. The pheromone is passed through the colony by the mutual licking and it acts on the individual by affecting its internal hormone balance, in particular the relative levels of juvenile hormone and molting hormone. Since these colonies are small, there is sufficient pheromone produced that no other reproductives develop; it is not a proportional but a total suppression.

An intermediate situation is found in the higher termites. There, it will be remembered, the colonies may become very large, so large that secondary reproductives will appear at the periphery of the colony. In this case one goes beyond the limit of the inhibitor pheromone of the primary reproductives; it becomes so dilute that it loses its power at the outer edge of the colony. Therefore, here we do have a crude ratio: one pair of reproductives can inhibit a finite number of molting nymphs, and beyond that threshold number the next level is reached where two pairs of reproductives are tolerated. The lesson to be learned is that a simple inhibitor can become a means of establishing a ratio of differentiated types in these social organisms.

A basically similar, but somewhat more precise, situation is found in the control of the ratios of the various worker classes. We have already discussed the selective advantage of such ratios in terms of Wilson's optimization theory, and here we want to concentrate on the basic principle of the mechanism. From the experimental studies of Light[13] it would appear that there are substances given off by soldier termites that inhibit worker nymphs from developing into soldiers. He showed this by two kinds of experiments. If the soldiers are all removed from a colony, new soldiers appear and soon the original soldier-worker ratio is restored. However, if the workers that are deprived of soldiers are fed a paste made up of mashed soldiers, there is a clear (but not total) inhibition of the formation of new soldiers at the succeeding molt. This case differs from that of reproductives in that the inhibitor pheromone is less potent and provides a balance between two castes, rather than one virtually eliminating the other. It could be assumed that the amount of inhibitor of a particular potency exuded per soldier is genetically determined, which in turn determines the ratio. The ratio could be altered by selection by making gene changes that alter the potency of the pheromone itself, the susceptibility of the workers to the pheromone, or the amount of pheromone produced per soldier.

There is one further factor involved, namely, that the substance is passed by the insects' licking one another. This trophallaxis is a rather specialized kind of diffusion, in which the molecules, instead of being propelled by their own kinetic energy, are carried by scurrying insects; it is in a literal sense a random walk.

From the examples we have given, clearly two related things can happen: first, in the higher termites a territory will be established for the inhibition beyond the limits of which the threshold of the inhibitor is too low to have any effect; second, there can be a relative dosage effect, so that two groups of workers can exist in a strict proportion. In this instance there is no spatial effect; all the termites freely diffuse and in this way there is a balance of castes. The simplicity of the case stems from the fact that both of these patterns, the proportions and the territories, can be effectively established in this way; there do not seem to be any subtle chemical control mechanisms. If one looks inside an individual and examines how the juvenile and the molting hormones affect the differentiation into a soldier or a reproductive, the chemical complexity of the case will increase vastly, as will the precision of the pattern developed. One need only look at the extreme differences in morphology between worker and soldier in some termites, and how each is perfectly patterned and proportioned. But we are anticipating, for we still are at the level above multicellular organisms.

Social insects also illustrate another very important method of producing a balanced, proportional division of labor. This is by having organisms of different ages together at one time. The best illustration is found in honeybees, although the same phenomenon is also found in other social insects. The life span of a worker bee is about 24 days and during that time there are certain obvious changes in her glandular structure. It is also clear from a meticulous study by Lindauer[14] that only very young workers clean out cells, middle-aged workers tend the brood and build combs, and foraging and dance-following are done entirely by bees in the last 4 days of their life. If all the workers were born at the same time, there would, of course, be no division of labor at any one moment, but because their appearance is staggered continuously throughout the warm months, at any one time there is a relatively stable proportion of chars, nurses, comb builders, and foragers. Lindauer did a most interesting experiment in which he managed to have all the older foraging bees leave a hive, and then substituted a new hive in its place. In this way he made two colonies out of one: one consisted entirely of foragers and the other entirely of housekeepers. The foragers' hive had no difficulty, for even though there was a bit of a domestic mess, food kept coming in. The inside of the other hive was spotless, but most of the workers starved to death; only at the last minute did some of them age sufficiently to begin bringing in food. From this experiment it is clear that not only is there a division of labor due to this temporal differentiation of worker bees of different ages, but it is clearly vital to the well-being of the colony.

As a footnote to this temporal division of labor, Wilson[15] points out that in ants both kinds of division of labor occur simultaneously. Each worker class or caste goes through a temporal change similar to those of the honeybee: nursing, nest work, and finally foraging. This may be true of both major and minor workers, while soldiers begin as nurses and end as defenders of the colony (Fig. 50). At any one moment, there is a balance between nursing, nest work, foraging, and defense.

The principal element in these examples of social insects is that the individuals are not physically connected. I would now like to consider another case in which this is at least functionally the case. It is an example on a much lower level of organization, but because of the virtual physical separation of parts, it is especially easy to study and analyze the methods of the chemical control of pattern.

The example is a growing mycelium of a soil fungus. The branching hyphal tip contains the main part of the protoplasm and the posterior portions are highly vacuolate; in this sense these hyphal tips act as separate individuals. Each time an advancing hypha branches, the branching is functionally equivalent to a cell division; in fact, in some species it is accompanied by a regular nuclear division and cross-wall formation.[16] It is well

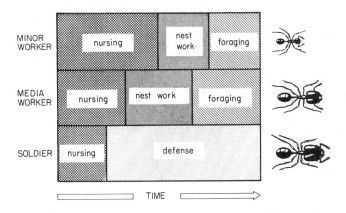

Fig. 50. The principal work periods traversed in the life spans of three worker subcastes of a generalized polymorphic ant species. The work periods are those periods in which the indicated task is the one most frequently performed; other tasks may be performed, but less often. The forms of the castes and the sequences of work periods within each caste are based on real species, but the precise durations of the periods are imaginary. In this case each of the eight periods, the (arbitrary) total number of periods encountered in all three castes together, is treated as a separate "caste." The optimal mix can be evolved by varying both the relative numbers in each subcaste and the relative time spent in each work period.   [From E. O. Wilson, *American Naturalist*, 102, 41–66 (1968).]

known that these advancing hyphal tips, either in the soil or in a culture dish, do not form a random pattern, but seem to make optimal use of the space available (Fig. 16). In other words, the hyphae are equidistant from one another and thus they can presumably sop up, in the most effective fashion, any soluble nutrient that is uniformly distributed.

The basis for this distribution is a diffusible growth inhibitor. One of the easiest ways of demonstrating it is to put a group of spores of a mold in one spot; the sprouting hyphal tips will grow away from the cluster of spores in a negative "group effect," to use the terminology applied to *Fucus* eggs. Also it can be shown by time-lapse photography that when a hyphal tip approaches another hypha either its growth slows down and stops, or it turns away and grows out in another direction.

Since we presume that the selective advantage of this self-inhibiting pattern is to consume the food available with maximum efficiency, one would expect an even spacing between all parts of the hyphae (Fig. 16). There is also some evidence that the hyphae will grow toward certain foods and it has been assumed that there must be a positive chemotaxis in these food-seeking hyphal tips. This may well be the case in some instances, but, as mentioned previously, D. R. Stadler[17] discovered that the tendency for the hyphal tips of *Mucor* to be attracted to turnip juice was due to a further inhibition rather than an attraction. The turnip juice apparently produces a substance that inhibits (or eliminates) the original inhibitor. In this way an especially steep gradient of the original inhibitor will be formed toward the turnip juice, with the result that the hyphae will grow straight for it.

We see in this instance a pattern produced by an inhibition that diffuses across considerable gaps between hyphae. The hyphae do not run about; they are quite motionless except for the tip growth, and this is the basis of the regular pattern. As far as chemical signals go, we can now add attractants and inactivating substances to our original inhibitor, all of which affect the pattern.

These molds live near the surface of soil or some other substratum, and therefore it is not surprising to find volatile attractants and repellents. This was first proposed years ago by Burgeff, but the first convincing evidence came from the work of Banbury, and now has been pursued by others.[18] *Mucor* exists in heterothallic strains and if the opposite sexes are placed on small agar blocks separated by an air space, the hyphae will bend toward one another and fuse (Fig. 51). If, on the other hand, two blocks both inhabited by the same sex are placed in close proximity, the hyphae will bend away from one another. Plempel suggests that these reactions are regulated by two volatile hormones, but the evidence on this point is not yet entirely satisfactory.

There are also many examples of attraction and fusion of hyphae in a liquid environment; it is a common phenomenon among molds. It would appear that the equidistant spacing is reserved for feeding, while the mutual attraction is reserved for reproduction leading to the exchange or transfer of nuclei. Attraction is also involved in another aspect of reproduction: the formation of compound fruiting bodies, a matter that we shall consider in the next section on the development of multicellular organisms.

A third and last example of organisms that communicate at a distance is the separate cell stage in the early development of the cellular slime molds. Part of the strategy in pursuing this case and that of soil mycelia of fungi is that both these organisms have a later stage in which the cells come together in close contact; in the first part of their life histories they coordinate by action at a distance and therefore serve as models for the social level, and in the second part they are proper multicellular organisms.

The main points concerning the mechanism of aggregation have already been described and discussed in detail. After depleting the food supply, all the amoebae, which are separate and evenly distributed over the substratum, simultaneously begin to secrete more acrasin and become more sensitive to it. As a result, a series of centers are formed and the amoebae stream into these centers by chemotaxis. The position of the centers or the pattern of a field of centers is nonrandom, and the reader is reminded that there appear to be three reasons for this: (1) each

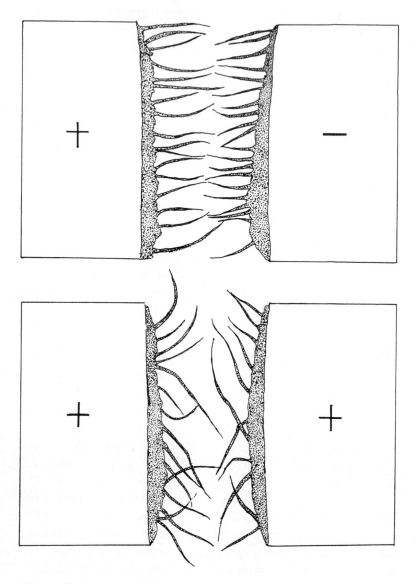

Fig. 51. Orientation of hyphae (Zygophores) of different sexes of *Mucor*, grown on agar blocks. Opposite sexes attract one another and the same sexes repel one another across the air gap between the blocks. [Drawings based on photographs by G. H. Banbury, *Journal of Experimental Biology, 6* (1935), Plate 6.]

center produces an inhibitor that prevents other centers from forming in its vicinity; (2) an acrasin gradient will appear around each center, and if a secondary center is in the midst of such a gradient it will be eliminated; (3) once the amoebae start going toward a center, their distribution becomes nonuniform, and this contributes to the ultimate nonrandomness. As we said, it is known that all three of the factors operate, so there is triple assurance that the nonrandom distribution will prevail. We also made the point that the territory size remained constant over a wide range of amoeba densities and from this made the assumption that this particular kind of distribution of centers, and therefore of fruiting bodies, must be adaptive for reasons of optimal spore dispersal.

The object of this example is to stress the great variability of the pattern of center formation (Fig. 39). Even though the territory size is roughly constant under certain conditions, the variation in both the size of any one territory (and hence of the fruiting body) and the actual pattern of the centers over a large area is considerable. One can only suppose that despite the three different ways of assuring some constancy in territory size, together they give only a very approximate result, one that we assume is adequate from an adaptive point of view.

Finally, in concluding this section, which involves patterns of living units that are separate in space, it is clear that the control signals are simple, often involving an inhibitory substance, and the patterns that result are approximate and imprecise.

2. *Patterns in multicellular organisms.* Let me begin by following through with two of the examples given in the previous section. Both soil fungi and cellular slime molds go from a "separate-cell" stage to a proper multicellular stage and produce a complex pattern. It is useful to compare, in both cases, the first stage with the second.

In the case of fungi, after the feeding is completed, or when the environmental conditions are favorable, the protoplasm in the mycelium streams to central points. At these points new hyphae form, but there is essentially no intake of energy. The new hyphae, which are gorged with the incoming protoplasm, no longer show the tendency (described earlier) to grow away from other hyphae; on the contrary, they cling to one another to form a mass of filaments bound together as in a fascicle. It is possible to describe many different kinds of compound fungus fruiting bodies, but let us concentrate on the common mushroom *Agaricus campestris*, which was first carefully studied long ago by de Bary.[19] The growing filaments produce a button made up of many small cells; in fact, the number of cells in the stalk of the button seems to approximate the number of cells in the mature stalk, which suggests that the final "growth" of the mushroom is entirely a matter of cell enlargement. In fact, if the wet-weight/dry-weight ratio of a "growing" mushroom is measured during the course of its size increase, it is found to remain constant, indicating that the increase in the enlargement of the cells is due to the rushing in of the protoplasm from the soil

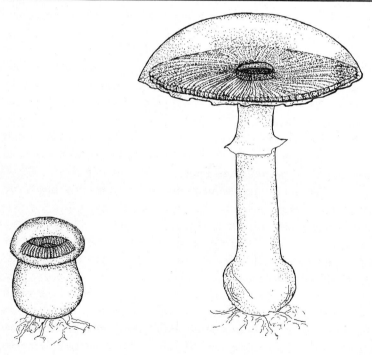

Fig. 52. A growing mushroom, showing the development of the gill region.

mycelia.[20] (In these fungi the cross walls of the cells have a pore or hole in the center, and both cytoplasm and nuclei can pass from one compartment to another.)

Rather early in the button stage, gill primordia are evident (Fig. 52). These expand along with the stalk and ultimately, as the cap opens up, the gills lie in perfect radial rows around the axis of the stalk, a feature familiar to us all. It had been known from some earlier work that if the cap with its gills was removed, the stalk would not elongate.[21] But the reason for this was unknown until Hawker[22] showed that if a small portion of gill was placed on one side of the decapitated stump, the parallel filaments on that side of the stalk elongated, producing a curvature. This experiment suggested that the gills might be giving off a hormone that stimulated cell elongation by the migration of protoplasm from the soil. This conjecture was fully verified by Urayama.[23] Hagimoto and Konishi[24] were able to extract this gill hormone and put it in agar blocks, which again would cause elongation if they were placed on the top of a decapitated stalk. From this we know how the elongation is controlled, but it leaves two major questions unanswered: how is the gill region, with its hormone-producing properties, established in the first place, and what controls the fine structure of the external form, for instance, the smooth rounded cap and the extraordinary precision of the gill clefts? Both of these features seem especially remarkable when one remembers that the inside of the cap or the gills is no more than a haphazard tangle of filaments (Fig. 53). This is the very thought that stimulated Gurwitsch[25] to suggest that somehow the development of pattern was controlled by a "field" analogous to an electromagnetic field.

But today we are not happy with the idea that we must depend upon some unknown physical force (such as Gurwitsch's mitogenetic rays) to describe pattern. We must somehow answer our two questions with familiar chemical and physical forces. Unfortunately, we have virtually no experimental data on either question, so all we can do is to make a few hypotheses. In the case of the localization of the gill primordium in the button stage, we might presume that there is some sort of diffusion pattern of

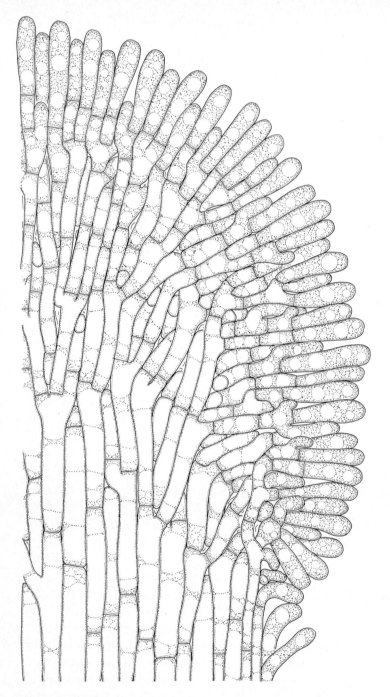

Fig. 53. Longitudinal section through the head of a fruiting body of the basidiomycete *Typhula gyrans* at the moment of its cessation of upgrowth. [From E. J. H. Corner, *Clavaria and Allied Genera* (Clarendon Press, Oxford, 1950), Fig. 47.]

a key morphogen in Turing's sense. The primordium is a torus-shaped structure inside the button (Fig. 52); it is easy to imagine some diffusion mechanism, perhaps involving an unstable substance, being distributed in such a pattern.

It is much harder to suggest an explanation for the external form of the mushroom and the fine structure of the gills. How do a tangle of filaments end up to form a relatively smooth surface (Fig. 53)? One thing is self-evident: the secret cannot lie in the individual filaments. They must somehow communicate with one another by the passing of chemical signals. These signals may involve (i) growth or, rather, elongation stimulators or inhibitors, or (ii) substances that affect the adhesion of the filaments, which in turn influence elongation, or both. The gill clefts are so precise that it is difficult to imagine that the pattern is the result of diffusion phenomena. One wonders whether there is not some macromolecular assembly of specific substances that somehow influences the shape at a key stage. The fact that some fungi have slit gills and others a series of fine pores (as in *Boletus*) might be accounted for by gene changes that affect this totally hypothetical macromolecule. This example is unsatisfactory because we lack experimental information, but let me stress two points. One is that there is no theoretical reason why the simple chemical explanations I propose might not be sufficient. Another is that, once we are at the level of the inside of a multicellular organism, then even though the organism may be constructed of a loose tangle of filaments, there are controls between the cells of far greater precision than anything we saw at the higher, social levels.

The second example that we began in the previous section on the social level also has a multicellular stage in its life history. Again in the cellular slime molds, we unfortunately know less about the key control events that occur in the cell mass, once the cells are collected together, than we do about the earlier aggregation stage where everything is spread out in two dimensions. The original center of the aggregate becomes the anterior tip of the slug; it retains its dominant role. Raper[26] showed that one could graft a number of tips onto one slug, and they divided the cells of the slug evenly among them (Fig. 54). The center began as a region of high acrasin emission, but this is not evident during the later stages of aggregation, where the large streams seem to emit as much acrasin as the center.[27] At the end of aggregation there is still an even secretion of acrasin, but as migration begins there is an emission gradient along the axis.[28] The interesting question arises whether this gradient is the result of the sorting out of cells that we know occurs at this point in the life cycle or whether the gradient preexists and in some way (perhaps by chemotaxis) guides the sorting out.

This brings us head-on to the central problem of the proportions of the stalk cells to the spore cells that are formed within the slug. Slug size varies enormously, since it depends upon how many amoebae enter an aggregation center. In *Dictyostelium discoideum* the ratio of spores to stalk cells remains proportionate over the entire size range.[29] In this same species it is also possible by means of histochemical techniques to identify the presumptive stalk and spore zones. The prestalk zone lies in the anterior end and is less than half of the total volume of the slug.

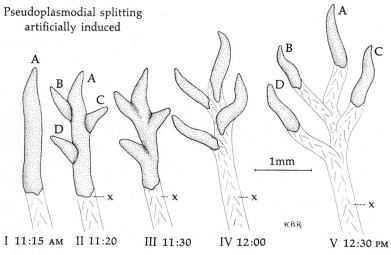

Pseudoplasmodial splitting artificially induced

I 11:15 AM   II 11:20   III 11:30   IV 12:00   V 12:30 PM

Fig. 54. Camera-lucida drawings of a migrating pseudoplasmodium of *Dictyostelium discoideum* showing the ability of grafted tips to induce the cells of the host to separate roughly equally between the four resulting tips. [From K. B. Raper, *Journal of the Elisha Mitchell Scientific Society,* 56, 241–282 (1940), Fig. 7.]

The appearance of these presumptive regions occurs after sorting out has taken place.

The main reason that this organism provides such a good example of spacing mechanism is that, as Raper[30] noted some years ago, if a slug is cut into portions, each fragment will form (given enough time) a perfect miniature fruiting body. Just as Driesch showed for hydroids, the slug has perfect powers of regulation and "the fate of a cell is a function of its position in the whole." In this case it can occur without any cell divisions; clearly the prestalk and prespore conditions are entirely reversible, without recourse to any cell cycles.[31] The proportions produced will be normal, although it does take longer for prestalk cells to turn into prespore cells than conversely; Gregg[32] showed that the prestalk cells need to go through a greater recapitulation of former steps. The final result is a remarkably exact restoration of the original spore-stalk cell proportions for each fragment. This is not quite the French-flag problem of Wolpert,[33] but each slug or isolated fraction of a slug is at least a perfect bicolor.

How is this accomplished in the slime molds? Regulation clearly can occur before or after the short interval of sorting out at the end of aggregation. If the cell mass is cut into segments early, presumably each piece sorts out separately; if it is done on a slug that has migrated some distance when the presumptive cells are already apparent, there is dedifferentiation. One might presume in the latter case that the sorting out itself produces a gradient of stalk and spore "tendencies," as I have called them, since they are so easily reversed. The hypothetical sequence would then be sorting out to give a gradient of differentiation tendencies that can then readjust if any cells are lost from either end. It is also quite possible that the acrasin-producing tip and the evidence for an acrasin gradient along the axis of the slug play a part; the gradient of acrasin emission certainly parallels that of the stalk-forming tendency, and the two could indeed be related.[34] (This hypothesis is especially intriguing now that we know that high concentrations of acrasin will directly promote stalk-cell formation in cells that have not aggregated.[35])

But a gradient is not enough to account for the strict proportionality between the prestalk and prespore zones. If one assumes that somehow a threshold in the gradient is responsible, one still is left with many problems. This would mean that, assuming no change in sensitivity of the cells to the graded morphogen, the absolute concentration of the morphogen must be the same at the division line for both gigantic and minute slugs, spanning a size range of three orders of magnitude. One interesting observation comes from comparing round, fat slugs with long, thin ones; the lengths of their prestalk and prespore zones do not show a proportional relation, only their volumes do so.[36] From this it is clear that somehow the posterior cells must "know" how many anterior cells there are and vice versa; the crucial information is cell number, not distances. This means that there have to be directional or polar messages and that these messages carry information about the quantity of protoplasm, not the distance traveled. We do know that there are a few especially fast-moving cells which continually move forward during migration, and a similar number that lag behind. Either one of these (or both) could carry such directional information in the form of some key substance, the quantity of which reflects the total number of cells in the end from which the cell came. It is, of course, equally possible that a morphogen that reflects the number of the cells could move in a polar fashion in the slug by some other means. In any event, whatever the detailed mechanism, it has to involve a communication between parts, in a way that has certain properties: (i) there is a polar movement of the key signal (or signals), (ii) the signal has the information of volume or cell number, and (iii) there is a response to the signal in the form of changing the differentiation to the correct proportions. This signal-response system could operate in a number of ways. For instance, the signal could be a positive stimulus, such as a specific substrate, for a reaction occurring at the tip, and the intensity of the reaction, hence the proportions, would be governed by how much signal appeared at the tip, reflecting the total volume of the posterior end. Or the signal could be an inhibitor, and since it moves in a polar fashion it would accumulate at one end. Again the quantity

242

of the inhibitor could affect the size of the respective prestalk and prespore zones. Not only is this kind of system capable of producing the initial stable proportions, but if it is grossly disturbed (for instance, by cutting it into pieces), it can recover and reform new, similar stable states in each piece.

It is useful to compare the control of proportions in cellular slime molds with the situation we have already described for insect castes, for in the latter case we were able to see a very simple explanation in the inhibitory properties of pheromones. First, there is a striking parallel in the fixity of the proportions. As we discussed, Wilson has effectively argued for an optimization of the caste ratios to explain their adaptive significance. Since stalk cells die and therefore are expensive in terms of reproductive cost, we can assume that the spore-stalk ratios in cellular slime molds must also be reaching an optimum, in this case for the effective dispersal of spores. The big difference between proportions in slime molds and in social insects is that in one the units producing the inhibitor are attached physically to one another and have a polarity of movement, while in the other the units are separate individuals that touch one another sporadically and thereby transmit the inhibitor. There is no polar organization in the termite colonies in the same sense that we find it in the slime mold.

The conclusion we may draw from this analogy is that cell systems could theoretically also produce proportions by the quantitative control of hormones and inductors. What we are doing in this discussion is to try to identify all the factors or parameters that could be responsible for proportionate patterns, and clearly the amounts of inhibitors or stimulators are important, but polarity, or direction of movement, is equally so. There is also the possibility that differential cell adhesion plays an important part.

Let us now turn to higher plants. We have already discussed the fact that both phyllotaxis and apical dominance can be controlled by chemical substances. In some cases we have seen that well-known plant hormones are involved; in other cases, such as phyllotaxis, the nature of the morphogen is not known. Again in these instances we have the possibility of inhibitors or stimulators, and even though it is not favored at the moment, there is also the possibility of competition for key metabolites as a means of producing pattern. The branch pattern produced by the hormone-controlled apical dominance is rather crude and variable, while the control of phyllotaxis, which works at a smaller, finer level, is remarkably precise. Once more we see an illustration of the rough correlation that the smaller the pattern the less variable it is.

Earlier we pointed out that though hormones, such as auxin, have profound effects on the morphology of the plant, there are more deep-seated tissue differences that are of even greater importance. The fact that, for instance, the root responds to certain levels of auxin by an inhibition of growth, while the shoot growth is stimulated by similar concentrations, shows that there is a fundamental difference in the two tissues. How did it arise? The answer lies in events that take place in the very early development of the fertilized egg. The localization of substances that give rise to root and shoot occurs during the first few cleavages after fertilization. The mechanism is not known, but there are numerous possibilities. There may be an initial uneven distribution of key cytoplasmic components in the egg, or following fertilization there could be a segregation, a migration of components into different parts (such as occurs in some mosaic animal eggs). Then the cleavage will compartmentalize these regions so that potential root cells will begin their development with different cytoplasms from the potential shoot cells.

This kind of explanation, namely, the localization of substances at an early stage of development, need not be irreversible. Again this is clearly shown in plants in which one cell from a shoot will produce a whole plant with both root and shoot. Also it is possible, in a much simpler experiment, to have a stem cutting give rise to roots, or a slice of a root (for example, a carrot slice) give rise to new shoots. In these latter cases the new tissue forms from a few cells that reverse their original direction of differentiation and begin a new one. This kind of phenomenon seems to involve a restoration of a balance between a certain amount of shoot tissue and a certain amount of root tissue. This balance is clearly not held in this case (or in normal development) by any rigid maintenance of the original distribution of cytoplasmic factors, but by the balance of chemical messengers that are given off by the two tissues. As in the maintenance of proportions in cellular slime molds, there is a continual communication between parts so that the amounts of root and of shoot tissue bear a constant (allometric) relation to each other.

In higher plants an additional phenomenon can be seen. As Goethe appreciated long ago (while contemplating plants in the old botanic garden at Padua), a flower is a modified stem. It is known, although the details have been elusive, that there is a flowering hormone responsible for this transformation. Through a phytochrome reaction, the day length is recorded, and if the plant is a "short-day" plant, then as the day length decreases in the fall the flowering hormone will, at a critical point, be released. In some way that is not yet understood it gathers in the tips of the shoots, and those shoots become transformed. Instead of producing whorls of leaves, the shoot tip now produces whorls of modified leaves: sepals, petals, stamens, and carpels. Clearly, all the cells of a growing tip had the potentiality of producing either vegetative leaves or reproductive flowers. The question of which of these two pathways the differentiation takes is entirely determined by the presence or absence of the flowering hormone. Once initiated, a flower develops, even if it is removed from any possible further source of hormone by placing it on a simple culture medium.

There are many parallels between higher plants and colonial hydroids, a fact of particular interest because plants build their shape with growth patterns of rigid, hard-walled cells, while hydroids have soft cells without cell walls, little or no evidence of growth zones, and a remarkable ability of the cells to move both in sheets and individually. In their overall shape they have branching patterns and even "phyllotactic" arrangements. Furthermore, like plants, hydroids can have either vegetative branches (feeding polyps) or reproductive branches (reproductive polyps).

In some colonial hydroids there is an even greater division of labor. For instance, *Hydractinea* has protective polyps in addition to feeding and reproductive ones. This kind of polymorphism reaches its greatest elaboration in the siphonophores, a group of free-swimming hydroids. They have, all in one organism, and all attached to one gastrovascular canal, a combination of different kinds of branches, which derive from both the polyp and the medusa stages. These modified "persons" in the colony will include, besides the feeding and reproductive polyps, different kinds of protective persons, swimming bells, and a float. In these very complex cases it is especially hard to sort out the mechanism of how the pattern and the division of labor are formed, whereas in the simpler hydroids it is relatively easy. Since there is no reason to believe that their methods of development differ except in terms of complexity, the simpler case will be instructive.

One might assume that the mechanism is essentially the same as in the flowering of plants: owing to a chemical difference in an expanding branch tip, the reproductive condition is initiated instead of the feeding polyp. Braverman[37] has shown that in a colony of the hydroid *Podocoryne* the reproductive polyps form in the center of the concentric colony, while the feeding polyps are at the outside edge (Fig. 55). It is not known what factors are responsible, but certainly there must be numerous chemical differences between the center of the colony and the edge. It was shown long ago by Peebles[38] that if the tip of a feeding or a protective polyp was cut off, the polyp regenerated back to its original state: a feeding polyp always regenerated to a feeding polyp, and a protective polyp stump always gave rise to a new protective polyp. As with the flower primordium, once it is started it is apparently fixed.

Comparable with the branching of plants (both fungi and higher plants), the pattern of branching in hydroids is exceedingly variable.[39] In neither case do we understand the mechanism, but again it is fairly clear that the pattern, however irregular, is not brought about by intrinsic unfolding of elements within the protoplasm, but by a communication between parts. Somehow

246

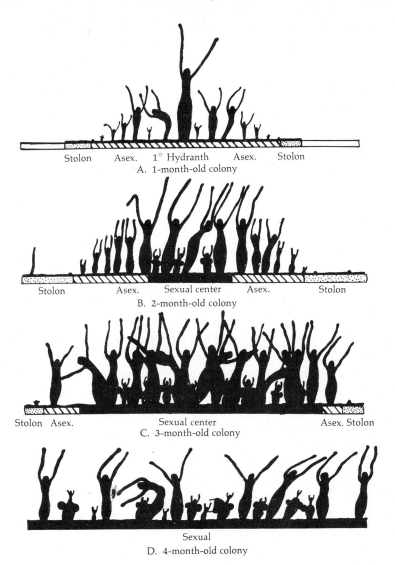

Fig. 55. Colonies of *Podocoryne* cut through the center and at right angles to the slide. For clarity, only two tentacles are shown on each zooid, and the size of the zooids is exaggerated. The relative size of the hydranths is correct. Shadings show the division of the colony into sexual, asexual, and stolon areas. [From M. M. Braverman, *Journal of Embryology and Experimental Morphology*, *11*, 245 (1963), Fig. 4.]

one branch point prevents another from occurring within a certain distance, and presumably it must do this by the release of one or more key morphogens.

It is only in the past few years that it has become evident that hydroids do not have meristems like plants, but that the cells divide fairly uniformly over the entire organism, and the change in form is due to the movement of individual cells and sheets of cells.[40] One of the most dramatic examples of this is the work on *Clytia* of Hale.[41] As this colony grows, new polyps appear at one end of the colony, branching off an advancing stolon while at the other end of the colony there is a regression of the polyps (Fig. 36). It has been known for some time that hydroids do regress on a cyclical basis, and this has been considered a case of controlled senescence. But Hale showed that the cells in these old hydranths do not senesce, but move up the gastrovascular canal and become part of the new hydranths and the advancing edge. The cells seem to go through cycles of adhesive and nonadhesive stages, so that at any one time in a colony new polyps are continually being formed and old ones are continually being disassembled. Since at one time the colony will have a wide spectrum of polyps of different ages, this building-disassembly cycle gives a kind of temporal differentiation akin to what is found in social insects. The only difference is that we have no clue to what the adaptive advantage might be for the hydroid colony.

Evidence for chemical communication between parts is also well known in the development of other animals, including vertebrates. There is a large amount of evidence for inductors, key substances that serve as chemical signals between one part of an embryo and another.[42] This began with the work of Spemann on the organizer, and has ramified into innumerable studies on many developing systems. Unlike plant hormones, these inductors seem generally to operate by contact; one tissue has to actually touch the other tissue in order for the subsequent development to occur. Because of this requirement of cell contact, the distribution of the inductor may be more carefully controlled, but in turn the question of what cells adhere to one another becomes a more important and critical issue. The rigid cells, the growth zones, and the rapid polar movement of hormones found in plants are now replaced by cell movement, specific cell adhesion, and the passage of chemical messengers from cell to cell. These really do represent radically different mechanical means of achieving similar goals.

Once an inductor acts, as in the case of the primary inductor of the dorsal lip of an amphibian blastopore, all sorts of detailed patterns emerge. As with roots and shoots, the pattern lies not in the inductor, but largely in the stimulated tissue, although there is evidence in some cases of regional differences in the inductors. In the case of the main axis of the embryo, one has first a setting up of a rough pattern which involves the blocking out of the direction of the main axis, the formation of the neural tube, and the subsequent formation of the mesoderm and the notochord. We do not know the details of the chemical sequence of events, but some general principles are clear. Each subsequent step following the initial action of the primary inductor involves interactions of parts. There are numerous secondary inductions, such as the neural-fold tube inducing the notochord, that follow in a sequence one after another. Instead of achieving a pattern in one step, as happens in the spore-stalk cell proportions of the cellular slime molds, they constitute a series of pattern steps in which one is a subunit of another. In vertebrates, these steps involve not only induction by cell contact, but cell movement, changes in the adhesive properties of cells in specific areas in the embryo, and cell proliferation. Because a vertebrate is so complex, its development is correspondingly difficult to unravel, but in principle there is every reason to think that the mechanism found in cellular slime molds and colonial hydroids is quite sufficient for an understanding of the development of a vertebrate. That is to say, not only do inductors and hormones communicate the direction of differentiation, but by feedback systems it is possible to produce a consistently proportioned pattern.

There is one intriguing and important exception to this view that pattern within multicellular organisms can be interpreted in terms of such chemical control loops. It is a sobering experience, for instance, to see the precision and symmetry of the pattern of the network of neurons in the eye of an insect. How is such a detailed lattice established by growing and moving nerve cells in development? The fact that it is consistently proportioned is all too obvious when one considers the complex functioning of the entire nervous system and that any defects will produce behavioral anomalies. There are many facts that must be unearthed if we are ever to answer this question.

It has been known for a long time, largely through the work of Sperry and Stone,[43] that if the eye of a newt is removed and replaced in some unusual position (say upside down), new axons will grow out from the retina to the optic tectum of the brain and make their *original* connections. Since the eye is rotated 180°, every time the newt sees something above him he reaches below. Recently Jacobson[44] has followed this process in development and finds that there is a strict time when this system becomes rigid or determined in the manner described above. If the experiment is performed at a very early stage, the animal sees normally, but after a specific embryonic stage, one finds the upside-down result of Stone. Furthermore, just as Harrison and his students[45] had found for limb and ear placode polarity in similar experiments many years earlier, Jacobson found that the anteroposterior axis became determined first, next the dorsoventral (Harrison found the third, or mesolateral, axis to be determined last in his experiments). To give an example, if the eyes were switched and the left eye placed upside down in the right eye socket, then both the A-P axis and the D-V axis would be reversed. If this was done at the time the A-P axis only was determined, then after regeneration of the new axones the animal would properly distinguish between up and down, but invert front and back. This determination sequence appears to be in the brain, in the optic tectum itself.

There is much food for thought in these experiments. First, we have the problem of how the nerve fibers find their right connections even when they are in some totally new orientation. The physicochemical basis of this is not clear. It cannot be a series of chemotactic gradients; there are too many nerves for such an accomplishment. Perhaps there is a general chemotropic attractant and once the axon endings are near the optic tectum they have specific binding sites with specific cells. At the moment this is idle speculation. Gaze[46] does have some evidence that there is not always a one-axon-to-one-cell specificity, but small groups of axons and groups of cells combine. Clearly, the answer to this problem will be of great significance to our understanding of the nervous system in general. The second problem is how determination takes place in successive axes. Harrison speculated that there might be some sort of supracrystal, some three-dimensional lattice that goes through the kinds of stages of orientation found in the transition of a liquid melt to a true crystal. There is, unfortunately, no evidence for such a macro-self-assembly lattice, but at the same time there are no other models (and Harrison made this proposal in the 1940's!). Surely this is a phenomenon of major significance, yet it remains obscure simply because it so stubbornly defies explanation. There are some superficial resemblances between nerve patterns and the cortical structures of ciliates, but whether there are any deep-seated similarities is unknown. One might speculate that both self-assembly and specificity, are in some way related to the properties of macromolecules, but this does not say enough.

The mechanism of specificity in retinal regeneration is in itself an interesting problem. The classical assumption is that each kind of macromolecule is genetically determined, yet there are not enough genes to account for all the individual neurons. As Judith Johnson[47] has pointed out, a similar situation arises in the production of antibodies, and now it is thought by various workers that variations in antibody structure might be due to somatic mutations which are then selected (Burnet's "clonal selection"). Johnson makes the suggestion that a similar mechanism might be involved in the formation of the many neurons; in this way the great diversity and specificity would not need a one-neuron–one-gene relation. In such a hypothesis one would also require conditions that carry out the internal selection, which may involve chemical gradients, cell movements, and specific cell adhesiveness. In fact, Johnson interprets the successive fixation

251

of axes of Harrison and Jacobson in terms of two- (or three-) dimensional gradients.

In this discussion of pattern in multicellular organisms we have stressed the fact that there can be control mechanisms that involve specific substances, the polar transport of these substances, the localization of these substances, the regional response to these substances by the localization of another set of substances, the different kinds of possible responses, such as substances stimulating, inhibiting, or competing, and substances affecting permeability, adhesion, and movement of cells. Any one of these steps can in turn be affected by events taking place in the cell cytoplasm or, ultimately, by genetic effects. The more complex the process, the greater the variety of instructions (both nuclear and cytoplasmic) inherited from previous generations.

3. *Pattern in single-cell organisms.* We now want to examine life-cycle development that occurs within one cell membrane or cell-wall boundary. Let us begin with the most complex and interesting case of multinucleate algae. They consist of one large tube, often of elaborate structure and sculpturing, in which the protoplasm, containing many nuclei, wanders about freely. Of these, the most remarkable of all is *Acetabularia*, because, as Hämmerling[48] showed, all of its nuclei are condensed into one large nucleus in the rhizoid. It is almost as though it were designed especially for the experimental morphologist, although the more likely explanation is that because the nucleus is buried in the rhizoid, when the cap or stalk is destroyed by wave action in a storm, regeneration can occur with maximum efficiency.

From the early work of Hämmerling it was known that there were nuclear products that could exist in an enucleate stalk for a matter of months and still promote the regeneration of the cap. It is thought that this nuclear material might be stable *messenger* RNA, but whether this interpretation turns out to be correct or not, the remarkable fact is that this nuclear substance can promote the sculptured detail of the cap, and furthermore designate these details correctly for the caps of different species (Fig. 56). This is all the more remarkable when one remembers that the cytoplasm is moving around continually at considerable speed by cyclosis. How can this rapidly moving messenger promote a species-specific pattern in the cap?

We do not know the answers, but again one can make some reasonable comments. In the first place, not all the cytoplasm is moving; that nearest the cell wall is stationary. Protein synthesis is continually taking place in the cytoplasm, even in enucleate stalks. Presumably the stable *messenger* RNA can continue to designate specific proteins in the active ribosomes. These proteins, which could lodge in the stationary cytoplasm adjacent to the cell wall, might conceivably, by forming a self-assembly pattern of their own, influence the deposition pattern of the new wall that is the species-specific character of the cap. So in this case we are not just relying on inductors and cell-surface properties, but on macromolecular self-assembly. The pattern, as one would expect, is correspondingly precise and symmetrical. The closest parallel found in developing cell masses would be to have specific cell sites on the surface and a "quasicrystal" of cells, as one might argue for the alga *Pediastrum* (Fig. 37). But macromolecular assembly can occur only within cells, and a coenocyte such as *Acetabularia* is an especially large kind of "single cell."

Another kind of specialized single cell is a ciliate protozoan. Here there is evidence of an especially intricate surface fine

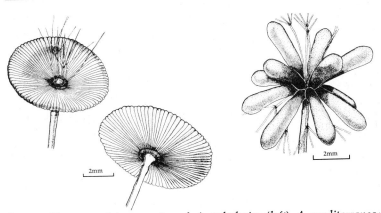

Fig. 56. The caps of two species of *Acetabularia*: (left) *A. mediterranea*; (right) *A. peniculus*. [From S. Puiseux-Dao, *Acetabularia and Cell Biology*, (Springer-Verlag, NewYork, 1970), Figs. 3 and 7.]

253

structure. As we have already noted, much of it can be passed on directly during binary fission to the daughter cells, but part of the structure will be reformed *de novo* after each fission. In suctorians it would appear that the entire fine structure is reformed in each generation (Fig. 35). Thus far we have stressed the quasicrystalline nature of the surface and have suggested that here also macromolecular self-assembly may play a role. But we did also point out that there are curious cortical movements that are totally unaccounted for by any kind of self-assembly hypothesis.

We can now add two further sets of facts that also support the idea that the ciliate surface is a combination of precise, crystallike events, and some more general interactions that remind us of multicellular systems. Of the former, Tartar[49] pointed out that the ciliate surface appears to be made up of basic units of set size. The basal body itself fits into this category, as do the collections of basal bodies or basal plates. During the formation of these basal plates in *Euplotes* they first expand to a set size and then separate to take up their appropriate positions on the ventral surface (Fig. 34). This example shows that basal bodies and aggregates of them act as "unit cells" analogous to the sense of the crystallographer; they form a basic, rigid building block.

But these units together form complex structures, most especially the oral groove and the whole pattern of the rows or kineties on the surface. The remarkable thing is that these structures as a whole are balanced and proportioned, so that if any alteration is made by surgery or by normal cell division, or by starvation, the proportions are restored, as de Terra[50] showed in *Stentor*, by the reorganization of the oral groove. The balance between the size of the groove and the size of the rest of the body is maintained within definable limits. One might argue that here also, as in all proportionate development, natural selection has favored an optimal ratio. The optimization theory that Wilson developed for caste proportions in an ant colony probably has very general application to the fixed proportions achieved in the development of all organisms.

The cortex of ciliates has a number of remarkable properties.

V. Tartar[51] was able to show that a particular region of the surface of *Stentor* (the region of stripe contrast) was a region capable of inducing new oral structures (Fig. 57). By grafting a secondary stripe-contrast region onto an individual he was able to induce secondary mouth structures. Unfortunately, we do not understand the mechanism of these surface interactions any more than we understand the cortical movements. It is conceivable that substances are passed from one region to another, but there is no evidence. It is interesting to note that A. S. G. Curtis[52] found in the fertilized eggs of amphibians that a small bit of the cortex from a specific region at this single-cell stage had the properties of the dorsal lip of the blastopore. He could repeat Spemann's experiments by transplanting small pieces of egg cortex instead of waiting until multicellular stages and transplanting bits of tissue. It may well be that the cortex of cells is so organized that it can give off induction substances from restricted regions. All the meager evidence we have seems to support this conclusion.

There is another interesting example of fine sculpturing of a single cell that also shows the same two properties: rigidity in some basic unit, yet an ability to produce a proportionate overall structure. Diatoms consist of a protoplast surrounded by two silica valves that fit into one another like a pillbox cover and bottom (Fig. 58). It is possible by osmotic-shock methods to spring the protoplasts free of the shells and grow them indefinitely as naked protoplasts on silica-free medium. If at any time one adds silica, they immediately build new shells which are perfectly formed and proportioned. The same process occurs normally in conjugation, where the zygote builds a new, larger shell, discarding the original shells of the conjugants.

In this last case there is a great increase in the size of the new shell, and during the course of normal successive cell divisions there is a slow, progressive decrease in the size of the shell. The reason for the latter is that each daughter cell, after fission, has one of the valves, and invariably builds a new valve inside the old one; only pillbox bottoms are made after binary fission. On the surfaces of the cell there are many fine holes of complex sculpturing, and these so-called pits are of fixed size. Therefore, if a shell is smaller there are fewer pits; if it is larger there are

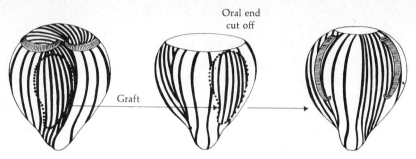

Oral end
cut off

Graft

Fig. 57. Induction of stripe contrast in *Stentor*. A section of fine striping has been grafted onto the flank of another individual so that now there are two regions where broad stripes lie next to narrow stripes, and both of these zones of stripe contrast induce new oral structures after the oral end of the host has been cut off. [After V. Tartar, *The Biology of Stentor* (Pergamon Press, London, 1961).]

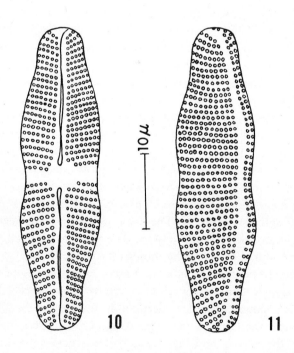

10μ

10   11

Fig. 58. Two views of a diatom (*Achnanthes coarctata*) showing the regular rows of pits that penetrate the silica shell. [From R. Patrick and C. W. Reimer, *The Diatoms of the United States* (Monographs of the Academy of Natural Sciences, No. 13, 1966), vol. 1, Plate 64, Figs. 10 and 11.]

more pits. This is the case even though the overall proportions of the valves will be normal. We may be certain that both the size of the pits and the shape of the whole shell are properties of the cell cortex. The invariance of the pit size would suggest that there is a rigid macromolecular structure in the cortex, but how this avoids conflicting with the mechanism that provides the overall proportions is an intriguing question.

As a final example, on a subcellular level, I would like to remind the reader of the remarkable development of T_4 phage. In the case of viruses, the building of specific proteins is entirely managed by tricking the host cell into manufacturing viral proteins instead of its own. The genes for many of these proteins have been identified, and the sequence of the construction of the tail, the tail fibers, and the head has been worked out in considerable detail (Fig. 31). It is of special interest that each of these three structures is manufactured separately, then the head is joined to the tail, and finally the tail fibers are added. Not all the assembly points are points of spontaneous self-assembly, as originally suggested, but they are assisted by enzymes or other factors; we have called this quasicrystalline form. Especially important is the fact that the developmental events first occur by making three units, and then these complex units are put together at the last stages. This has the sophistication of a modern automobile assembly line; it shows how even the relatively simple development of a virus has a cumulative set of sequences and subsequences.

The end result is a structure of almost crystalline perfection. Here the proportions do not seem to vary within a narrow range, but appear quite fixed. We assume that the reason for this increase in precision comes from the fact that molecular forces are responsible for the shape; these are bound to be more exact than the type of proportions that involve diffusion gradients and cell movements.

Summary and conclusions

From this survey of the control of patterns certain generalizations can be made. In the first place, it is obvious that organisms

257

of all sizes and complexities, even the social groupings of insects, develop into forms that have a controlled proportionality from one life cycle to the next. It is postulated that the proportions are established by natural selection to produce optimally efficient individuals (or social colonies) during certain key periods of their life cycles.

In large measure the developmental control of this proportionality is achieved by chemical signals. Where the living units are separate, as in social insects, certain stages in the life cycle of cellular slime molds, and various fungi, there are pheromones or other chemical messengers that act at a distance. The means of transport of the messenger is either being actively carried by the individual units (ants or cells) or diffusion.

In multicellular organisms the mechanisms of transport of the signals (hormones and inductions) include diffusion, cell movement, active transport, and polar movement. From the very fact that the cells are touching, the chemical messengers can be passed in a greater variety of ways, and the possibilities of control are thereby increased.

In the smaller level of single cells we begin to see form-making phenomena that could best be interpreted in terms of a combination of macromolecular interactions, active transport, and chemical signals diffusing over short distances. The cortex of the cell becomes an especially active site for such activities. Organelles within cells and viruses have an even greater precision in their shape and proportions, presumably because of an increased reliance on the combining of key macromolecules in forming what we have called a quasicrystal.

Note that the order of these three levels of organization is in a rough sense the reverse of that which occurred during evolution. This means that initially the first steps were made on the relatively precise, quasicrystalline level, and as there was a selection for a size increase, there was an automatic shift to the larger, higher-level control mechanisms. It is a trend of increasing action at a distance. The macromolecules fit together with weak bonds, but soon the aggregate of living material becomes so large that there must be other means of transmitting chemical signals; this is where we see, among other mechanisms, diffusion and even the transport of the signals by moving protoplasm. The system of intercommunication to achieve the control of the pattern is therefore strongly affected by size.

Since organelles lie within cells, which lie within multicellular organisms, which lie within social groupings, there are control cycles within control cycles. Each level is entrained by the level above it, and the highest level is the life cycle (or colony cycle in social insects). The more complex the organism, the greater the amount of information it carries in the single-cell stage that begins its life cycle. All of the information it carries is encoded in the DNA, but, as I have repeatedly emphasized, some of this information is already expressed in the single-cell stage of the cycle: the existence of the cell itself is proof that this is so. So each cycle starts with DNA ready to give off developmental messages, surrounded by protoplasm that is the result of a plethora of previous messages. The combination of the DNA and the DNA-derived information is sufficient to build a complex life cycle. This information produces a series of developmental changes involving the carefully controlled production of substances, the timing of their production, and their localization. This is all managed in such a way that each development produces an efficiently proportioned individual (or society), consistent from one life cycle to the next. Once a replication system appeared in the origin of life, natural selection has favored a series of slow, steady-state evolutionary developments: the minimum-size, single-cell stages of each life cycle which give rise to rapid, successive, self-replacing, and constantly changing life-cycle developments. The cost of reproduction goes up with each new addition to the complexity of the life cycle, but the additions have nevertheless been favored by selection, for they provide advantages in successfully reproducing offspring and contributing individuals of their own genetic kind to future generations.

Notes

1. *Introduction*

    1. C. G. Hempel and P. Oppenheim, *Philosophy of Science, 15,*   p. 2
135–175 (1948).

2. *Cycles and Natural Selection*

    1. L. E. R. Picken, *The Organization of Cells and Other Orga-*   p. 13
nisms (Clarendon Press, Oxford, 1960)

    2. H. Quastler, unpublished lecture; G. A. Sacher, *CIBA Founda-*   p. 16
tion Colloquia, 5, 115-133 (1959).

    3. P. B. Medawar, *The Uniqueness of the Individual* (Methuen,
London, 1957), pp. 17–70.

    4. G. C. Williams, *Evolution, 11,* 398–411 (1957).

    5. W. D. Hamilton, *Journal of Theoretical Biology, 12,* 12–45
(1966).

    6. J. M. Emlin, *Ecology, 51,* 588–601 (1970).

    7. R. J. Britten and E. H. Davidson, *Quarterly Review of Biology,*   p. 18
46, 111–133 (1971).

3. *Reproduction*

    1. J. T. Bonner, *Size and Cycle* (Princeton University Press,   p. 21
Princeton, N.J., 1965).

    2. For a review, see J. T. Bonner, *Journal of Paleontology, 42*   p. 22
(Part II of II), 1–15 (1968).

    3. S. M. Stanley *Evolution, 27,* 1–26 (1973).

    4. G. G. Simpson, *The Major Features of Evolution* (Columbia   p. 23
University Press, New York, 1953), p. 218.

    5. Bonner, *Journal of Paleontology.*   p. 24

    6. A similar curve has been previously published by C. E. Yar-   p. 25
wood, *American Naturalist, 90,* 97–102 (1956), Fig. 1.

    7. See R. H. Mac Arthur and E. O. Wilson, *The Theory of Island*   p. 27
Biogeography (Princeton University Press, Princeton, N.J., 1967).

    8. W. M. Schaffer, Ph.D. thesis, Princeton University (1972).   p. 28

    9. *Ibid.*; M. Gadgil and W. H. Bossert, *American Naturalist, 104,*
1–24 (1970).

    10. E. O. Wilson, *The Insect Societies* (Harvard University Press,   p. 29
Cambridge, Mass., 1971). For example, some of the largest colonies

are army ants, in which one family consists of more than 20 million individuals and will weigh a total of 20 Kg.

p. 30   11. D'Arcy Thompson, *On Growth and Form* (Cambridge University Press, Cambridge, England, 1942).

    12. J. T. Bonner, *Morphogenesis* (Princeton University Press, Princeton, N.J., 1952).

p. 32   13. I. I. Schmalhausen, *Factors of Evolution* (Blakiston, Philadelphia, 1949).

    14. R. Levins, *Evolution in Changing Environments* (Princeton University Press, Princeton, N.J., 1968).

4. *Inherited Variation*

p. 37   1. In J. A. V. Butler and D. Noble, eds., *Progress in Biophysics and Molecular Biology* (Pergamon Press, Oxford, 1971). vol. 25, pp. 145–190.

p. 42   2. F. Schnepf and R. M. Brown, Jr., in J. Reinert and H. Ursprung, eds., *Origin and Continuity of Cell Organelles* (Springer-Verlag, New York, 1971), pp. 299–322.

p. 43   3. For a review, see C. Fulton, *ibid.*, pp. 170–221.

    4. G. W. Grimes, *Journal of Cell Biology*, 57, 229–232 (1973).

p. 44   5. A. P. Mahowald, in *Origin and Continuity of Cell Organelles*, pp. 158–169.

p. 47   6. From a lecture by T. C. Schneirla. See also his *Army Ants* (W. H. Freeman, San Francisco, 1971), pp. 146–148.

    7. E. O. Wilson, *The Insect Societies* (Harvard University Press, Cambridge, Mass, 1971), p. 221.

    8. *Ibid.*, pp. 224–232.

p. 48   9. J. H. Sudd, *Discovery*, London (June 1963), pp. 15–19.

p. 52   10. In J. M. Allen, ed., *The Nature of Biological Diversity* (McGraw-Hill, New York, 1963), pp. 165–221.

    11. M. Cohen, in O. J. Plescia and W. Braun, eds., *Nucleic Acids in Immunology*, (Springer-Verlag, New York, 1968), pp. 671–715.

    12. J. E. Ackert, *Genetics*, 1, 387–405 (1916).

p. 54   13. J. T. Bonner, *Size and Cycle* (Princeton University Press, Princeton, N.J., 1965).

    14. C. D. Darlington, *Evolution of Genetic Systems* (2nd ed.; Oliver and Boyd, Edinburgh, 1958), pp. 223–227.

p. 55   15. It has been suggested by E. C. Cox in *Nature New Biology*, 239, 133–134 (1972) that mutator genes may even possess the ability to control the rates of mutation of parts of the chromosomes.

16. Admirably discussed in G. R. de Beer, *Embryos and Ancestors*   p. 57
(3rd ed.; Clarendon Press, Oxford, 1958).
17. *Ibid.* This is the basis of de Beer's "clandestine evolution."
18. Wilson, *The Insect Societies.*   p. 60
19. de Beer, *Embryos and Ancestors.* See also the discussion of
"chains of steps" in Bonner, *Size and Cycle.*
20. L. L. Whyte, *Internal Factors in Evolution* (Braziller, New   p. 61
York, 1965).
21. R. J. Britten and E. H. Davidson, *Science, 165,* 349–357 (1969).
22. I. Franklin and R. C. Lewontin, *Genetics, 65,* 707–734 (1970).   p. 62
23. An excellent, very brief summary of this problem is given by   p. 63
E. G. Leigh, Jr., *Adaptation and Diversity* (Freeman, Cooper, San Francisco, 1971), p. 242.

5. *The Levels of Complexity*
1. For a clear, brief summary of the earlier hypotheses on the ori-   p. 65
gin of life, and an interesting, unorthodox one of his own, see A. F.
Cairns-Smith, *The Life Puzzle* (Oliver and Boyd, Edinburgh, 1971).
2. For a review see S. Spiegelman, *Quarterly Reviews of Biophys-*   p. 66
ics, 4, 213–253 (1971).
3. L. E. Orgel, *Journal of Molecular Biology, 38,* 381–393 (1968).
4. S. L. Miller, *Science, 117,* 528–529 (1953).
5. S. Spiegelman, *Quarterly Reviews of Biophysics, 4,* 213–253
(1971).
6. From this it is obvious that evolution will therefore favor an in-   p. 68
crease in the size of the genome. See S. Ohno, *Evolution by Gene Dup-*
lication (Springer-Verlag, New York, 1970).
7. F. H. C. Crick, *Journal of Molecular Biology, 38,* 367–379   p. 69
(1968).
8. Ohno, *Evolution by Gene Duplication.*   p. 70
9. *Ibid.*   p. 72
10. S. Spiegelman, *Quarterly Reviews of Biophysics, 4,* 213–253   p. 73
(1971).
11. See J. D. Watson, *The Molecular Biology of the Gene* (2nd ed.;   p. 76
Benjamin, New York, 1970), for a good discussion of the chemical con-
tents of a cell of *E. coli.*
12. A. B. Pardee, *Symposia of the Society for General Microbi-*   p. 78
ology, 11, 19–40 (1961).
13. J. D. Bernal, in M. Florkin, ed., *Aspects of the Origin of Life*   p. 79
(Pergamon Press, Oxford, 1960), pp. 155–169.

p. 86   14. J. Adler, *Science*, *166*, 1588–1597 (1969).

p. 87   15. For a review of cellular slime mold chemotaxis, including the work on cylic AMP, see J. T. Bonner, *Annual Review of Microbiology*, *25*, 75–92 (1971).

16. P. Pan, E. H. Hall, and J. T. Bonner, *Nature*, *237*, 181–182 (1972).

17. E. W. Samuel, *Developmental Biology*, *3*, 317–335 (1961); also further unpublished observations in our laboratory.

p. 88   18. M. Dworkin, *Critical Reviews in Microbiology*, *2*, 435–452 (1972).

p. 90   19. D. R. Coman, *Archives of Pathology*, *29*, 220–228 (1940); M. J. Carlile, *Journal of General Microbiology*, *63*, 221–226 (1970); see also T. M. Konijn and J. L. Koevenig, *Mycologia*, *63*, 901–906 (1971).

p. 93   20. D. R. Stadler, *Journal of Cellular and Comparative Physiology*, *39*, 449–474 (1952); *Biological Bulletin*, *104*, 100–108 (1953).

p. 94   21. M. Aschner and J. Cronin-Kirsh, *Archiv für Mikrobiologie*, *74*, 308–314 (1970).

22. J. T. Bonner and F. E. Whitfield, *Biological Bulletin*, *128*, 51–57 (1965).

23. A. R. H. Buller, *Researches on Fungi*, vol. 1 (Longman, Green, London, 1909), pp. 47–78.

24. J. T. Bonner and M. R. Dodd, *Developmental Biology*, *5*, 344–361 (1962).

25. Unpublished observations of K. Zachariah and E. L. Martin in our laboratory.

p. 100   26. D'Arcy Thompson, *On Growth and Form*.

27. For example, M. Kimura and T. Ohta, *Theoretical Aspects of Population Genetics* (Princeton University Press, Princeton, N.J., 1971).

p. 101   28. J. R. Baker, *Nature*, *161*, 548–551; 587–589 (1948).

29. For a review see A Kühn, *Lectures on Developmental Physiology* (Springer-Verlag, New York, 1971), pp. 112–118.

p. 104   30. E. C. Dougherty *et al.*, eds., *The Lower Metazoa* (University of California Press, Berkeley, 1963); J. Hadzi, *The Evolution of the Metazoa* (Macmillan, New York, 1963).

p. 105   31. G. P. Bidder, *Quarterly Journal of Microscopic Science*, *67*, 293–323 (1923); *Proceedings of the Linnean Society of London*, *147*, 119–146 (1936–37).

p. 107   32. R. Rasmont, in D. Rudnick, eds.. *Regeneration* (20th Growth Symposium; Ronald Press, New York, 1962).

p. 108   33. E. O. Wilson, *The Insect Societies* (Harvard University Press,

Cambridge, Mass., 1971), p. 5.

34. *Ibid.*, pp. 4–5.   p. 108

35. *Ibid.*, pp. 146–147.   p. 110

36. *Ibid.*, pp. 327–335. In these pages Wilson has an excellent re-   p. 111
view and critical discussion of Hamilton's work.

37. *Ibid.*, p. 328.

38. *Ibid.*, pp. 130–134.   p. 112

39. *Ibid.*, pp. 120–135.   p. 114

40. *Ibid.*, p. 119.   p. 116

41. J. H. Cook, *Symposia of the Zoological Society of London*, 14,   p. 117
181–218 (1965).

42. F. F. Darling, *Birds, Flocks, and Breeding Cycles* (Cambridge
University Press, Cambridge, England, 1938).

43. A. Murie, *The Wolves of Mt. McKinley* (United States Gov-
ernment Printing Office, Washington, D.C., 1944).

44. Wilson, *The Insect Societies*, p. 10.   p. 119

45. *Ibid.*, p. 299.

46. R. A. Mac Arthur, *Geographical Ecology* (Harper and Row,
New York, 1972), points out that aggression between competing species is a secondary or more advanced condition that is added to the
primary competition for resources. Apparently there can be no general rules concerning the origin of aggression.

47. An exception in insects is the work of W. E. Kerr on castes in
the stingless bees which are genetically determined (see Wilson, *The
Insect Societies*, pp. 177–179). Also it should be mentioned that little
is known of the way in which dominance and peck order are determined in vertebrates.

48. Wilson, *The Insect Societies*, review on pp. 139–146.   p. 120

49. *Ibid.*, review on pp. 341–348.

50. M. Lindauer, *Communication among Social Bees* (Harvard   p. 123
University Press, Cambridge, Mass., 1961).

6. *The Synthesis of Substances*

1. *Biochemistry and Morphogenesis* (Cambridge University Press,   p. 131
Cambridge, England, 1942), pp. 505–531.

2. For example, J. D. Watson, *Molecular Biology of the Gene* (2nd   p. 135
ed.; W. A. Benjamin, New York, 1970); C. L. Markert and H. Ursprung, *Developmental Genetics* (Prentice Hall, Englewood Cliffs,
N.J., 1971).

3. A. Braun, *The Cancer Problem* (Columbia University Press,   p. 138
New York, 1969).

p. 138   4. Reviewed in Watson, *Molecular Biology of the Gene.*

5. J. M. Mitchison, *The Biology of the Cell Cycle* (Cambridge University Press, Cambridge, England, 1971).

p. 139   6. *Ibid.*, reviewed on pp. 169–180.

7. *Ibid.*, pp. 170–173.

8. R. W. Turkington, in I. L. Cameron, G. M. Padilla, and A. M. Zimmerman, eds., *Developmental Aspects of the Cell Cycle* (Academic Press, New York, 1971), pp. 315–355.

p. 140   9. Reviewed by P. C. Newell in *Essays in Biochemistry, 7*, 87–126 (1971).

10. C. K. Leach and J. M. Ashworth, *Journal of Molecular Biology, 68*, 35–48 (1972).

11. J. F. Varner, *Symposia of the Society for Experimental Biology, 25*, 197–206 (1971).

p. 141   12. N. Sueoka and T. Kano-Sueoka, *Progress in Nucleic Acid Research and Molecular Biology, 10*, 23–55 (1970).

p. 142   13. J. P. Changeux, *Scientific American, 212*, 36–45 (April 1963).

p. 145   14. Reviewed by A. B. Pardee in *The Harvey Lectures*, series 65, pp. 59–71 (1971).

15. See R. Rosen, *Dynamical System Theory in Biology* (Wiley Interscience, New York, 1970).

p. 146   16. Reviewed in Braun, *The Cancer Problem*, pp. 99–102.

17. F. Meins, Jr., *Developmental Biology, 24*, 287–300 (1971).

p. 147   18. Reviewed in F. Skoog and C. O. Miller, *Symposium of the Society for Experimental Biology, 11*, 118–131 (1957).

19. E. Hadorn, *Brookhaven Symposia in Biology, 18*, 148–161 (1965).

p. 148   20. E. Fauré-Fremiet and H. Mugard, *Comptes Rendus de l' Académie des Sciences, 227*, 1400 (1948).

21. Reviewed in Markert and Ursprung, *Developmental Genetics*, pp. 123–146.

p. 149   22. L. Goldstein and D. M. Prescott, *Journal of Cell Biology, 33*, 637–644 (1967); *ibid., 36*, 53–61 (1968).

23. For an excellent review of recent work see M. Hamburgh, *Theories of Differentiation* (Edward Arnold, London, 1971).

p. 150   24. K. R. Yamamoto and B. M. Alberts, *Proceedings of the National Academy of Sciences. U.S., 69*, 2105–2109 (1972).

25. S. Ohno, *Nature, 234*, 134–137 (1971).

p. 151   26. G. A. Robison, R. W. Butcher, and E. W. Sutherland, *Cyclic AMP* (Academic Press, New York, 1971).

27. I. Pastan and R. Perlman, *Science, 169,* 339–344 (1970).   p. 151

28. For a review see Sueoka and Kano-Sueoka, *Progress in Nucleic*   p. 152
Acid Research and Molecular Biology, p. 46.

29. E. O. Wilson, *The Insect Societies* (Harvard University Press,   p. 153
Cambridge, Mass., 1971): for ants, pp. 146–156; for bees and wasps,
pp. 171–181; for termites, pp. 188–196.

30. The work of W. E. Kerr is reviewed by Wilson, *ibid.,* pp. 177–
179.

31. For a review see Wilson, *ibid.,* pp. 188–194.   p. 154

32. C. M. Child, *Patterns and Problems of Development* (Univer-   p. 156
sity of Chicago Press, Chicago, Ill., 1941), pp. 413–421.

7. Timing

1. H. G. Callan, *Proceedings of the Royal Society,* London, ser. B,   p. 159
181, 19–41 (1942).

2. Reviewed by C. C. Markert and H. Ursprung, *Developmental
Genetics* (Prentice Hall, Englewood Cliffs, N.J., 1971). See also H.
Harris, *Nucleus and Cytoplasm* (2nd ed.; Clarendon Press, Oxford,
1970).

3. Reviewed by Markert and Ursprung, *Developmental Genetics.*

4. All the original work on the morphology and the grafting was   p. 160
done by J. Hämmerling, reviewed in S. Puiseux-Dao, *Acetabularia and
Cell Biology* (Logos Press, London, 1970).

5. Harris, *Nucleus and Cytoplasm.*

6. Reviewed by Markert and Ursprung, *Developmental Genetics.*

7. R. Goldschmidt, *Physiological Genetics* (McGraw-Hill, New
York, 1938).

8. E. B. Ford and J. S. Huxley, Roux's *Archiv für Entwicklungsmechanik des Organism, 117,* 67–79 (1929).

9. Reviewed in P. C. Newell, *Essays in Biochemistry, 7,* 87–126   p. 161
(1971).

10. *De Generatione Animalium,* trans. A. Platt (Clarendon Press,
Oxford, 1912).

11. A. Dalcq, *Form and Casualty in Early Development* (Cambridge
University Press, Cambridge, England, 1938).

12. C. H. Waddington, *Organisers and Genes* (Cambridge University Press, Cambridge, England, 1940); *The Strategy of the Genes*
(Allen & Unwin, London, 1957).

13. J. B. S. Haldane, *American Naturalist, 66,* 5–24 (1932).   p. 162

14. A. E. Boycott, C. Driver, S. C. Garstang, and F. M. Turner,

Philosophical Transactions of the Royal Society, London, ser. B, *219*, 51–130 (1930).

p. 162   15. H. Spurway, *Journal of Genetics*, *49*, 126–140 (1948).

16. J. Mandelstam, *Symposia of the Society for Experimental Biology*, *15*, 1–26 (1971).

p. 163   17. Reviewed by P. C. Newell, *Essays in Biochemistry*, *7*, 87–126 (1971).

18. *Ibid.*

19. J. M. Mitchison, *The Cell Cycle* (Cambridge University Press, Cambridge, England, 1971).

p. 165   20. M. Lindauer, *Communication among Social Bees* (Harvard University Press, Cambridge, Mass., 1961).

21. J. T. Bonner and M. J. Shaw, *Journal of Cellular and Comparative Physiology*, *50*, 145–154 (1957).

p. 166   22. S. Hendrichs, in A. Lang, ed., *Communication in Development* (28th Symposium, *Developmental Biology* Supplement; Academic Press, New York, 1969), pp. 227–243.

23. *Ibid.*

p. 167   24. B. C. Goodwin and M. H. Cohen, *Journal of Theoretical Biology*, *25*, 49–107 (1967).

25. R. H. Mac Arthur, *Geographical Ecology* (Harper and Row, New York, 1972).

p. 169   26. V. G. Bruce, *Genetics*, *70*, 537–548 (1972).

27. C. S. Pittendrigh, *Proceedings of the National Academy of Science, U.S.*, *58*, 1762–1767 (1967).

8. *The Localization of Substances*

p. 171   1. Reviewed in *The Science and Philosophy of the Organism* (Black, London, 1908), vol. 1, pp. 126–129.

p. 172   2. L. Wolpert, *Current Topics in Developmental Biology*, *6*, 183–224 (1971).

3. T. H. Morgan, *Regeneration* (Macmillan, New York, 1901).

p. 173   4. H. Spemann, *Embryonic Development and Induction* (Yale University Press, New Haven, Conn., 1938).

5. For a good discussion of Gurwitsch's work see L. von Bertalanffy and J. H. Woodger, *Modern Theories of Development* (Oxford University Press, London, 1933).

6. C. M. Child, *Patterns and Problems of Development* (University of Chicago Press, Chicago, Ill., 1941).

7. A. M. Turing, *Philosophical Transactions of the Royal Society*, London, ser. B, *237*, 37–72 (1952).

8. For an excellent discussion of what he calls "subcrystals" see   p. 174
A. G. Cairns-Smith, *The Life Puzzle* (Oliver and Boyd, Edinburgh,
1971), chap. 3.

9. Reviewed in W. B. Wood and R. S. Edgar, *Scientific American,*   p. 176
217, 65–74 (July 1967).

10. U. K. Laemmli, *Nature, 227,* 680–685 (1970); see also J. King   p. 177
and U. K. Laemmli, *Journal of Molecular Biology, 62,* 465–477 (1971).

11. Reviewed by C. Fulton in J. Reinert and H. Ursprung, eds.,   p. 178
Origin and Continuity of Cell Organelles (Springer-Verlag, New York,
1971), pp. 170–221.

12. T. M. Sonneborn, in J. M. Allen, ed., *The Nature of Biological*   p. 179
Diversity (McGraw-Hill, New York, 1963), pp. 165–221.

13. E. Fauré-Fremiet, *Bulletin Biologique de la France et la*   p. 180
Belgique, 79, 106–150 (1945). The importance of polarity in grafting
has been shown by numerous authors; see especially V. Tartar, *The*
Biology of Stentor (Pergamon Press, London, 1961).

14. Reviewed by T. M. Sonneborn, *Proceedings of the Royal So-*
ciety, London, ser. B, *176,* 347–366 (1970).

15. This is elegantly reviewed in D'Arcy Thompson *On Growth*   p. 181
and Form (Cambridge University Press, Cambridge, England, 1942).

16. E. Fauré-Fremiet, *Archives d'Anatomie Microscopique et de*
Morphologie Expérimentale, 42, 209–225 (1953).

17. S. J. Singer and G. L. Nicolson, *Science, 175,* 720–731 (1972).   p. 183

18. J. T. Bonner, *Journal of Morphology, 95,* 95–108 (1954).

19. Reviewed in A. Lwoff, *Problems of Morphogenesis in the Cili-*
ates (Wiley, New York, 1950).

20. G. W. Grimes, *Journal of Cell Biology, 57,* 229–239 (1973);
Genetical Research, 21, 57–66 (1973).

21. J. S. Huxley, *Quarterly Journal of Microscopical Science, 65,*   p. 186
293–322 (1921).

22. Reviewed by J. P. Trinkaus, *Cells and Organs* (Prentice Hall,
New York, 1969).

23. A recent review of M. S. Steinberg's ideas and experiments is
in the *Journal of Experimental Zoology, 173,* 395–434 (1970).

24. *Ibid.*   p. 187

25. Reviewed in J. T. Bonner, *Annual Review of Microbiology, 25,*
75–92 (1971). Special mention should be made of the work of I.
Takeuchi.

26. L. J. Hale, *Journal of Embryology and Experimental Morphol-*   p. 188
ogy, 12, 517–538 (1964).

27. J. G. Moner, Ph.D. thesis, Princeton University (1953).   p. 190

p. 190

28. B. M. Shaffer, *Advances in Morphogenesis*, 2, 109–182 (1962); 3, 301–322 (1964). For some most interesting recent experiments see H. Beug, F. E. Katz, and G. Gerisch. *Journal of Cell Biology*, 56, 647–658 (1973).

p. 191

29. N. Rashevsky, *Mathematical Biophysics* (rev. ed.; University of Chicago Press, Chicago, Ill., 1938).

30. P. A. Lawrence, *Symposia of the Society for Experimental Biology*, 25, 379–390 (1971). See also P. A. Lawrence, F. H. C. Crick, and M. Munro, *Journal of Cell Biology*, 11, 815–853 (1972).

31. J. T. Bonner, *Journal of Experimental Zoology*, 110, 259–271 (1949). More recent evidence from the work of Dr. P. Pan in our laboratory suggests that, while the secretion of acrasin in a cell mass may follow a gradient in *Dictyostelium*, the bound cyclic AMP in the cells appears at a uniform relatively high concentration in the anterior, prestalk cells, as compared to a lower uniform distribution in the posterior, prespore cells.

p. 192

32. B. M. Shaffer, *American Naturalist*, 91, 19–35 (1957).

33. F. H. C. Crick, *Symposia of the Society for Experimental Biology*, 25, 429–438 (1971).

34. B. C. Goodwin and M. H. Cohen, *Journal of Theoretical Biology*, 25, 49–107 (1969).

35. G. F. Sleggs, *Growth*, 3, 173–179 (1939).

p. 193

36. Reviewed in L. F. Jaffee, *Advances in Morphogenesis*, 7, 295–328 (1968).

p. 194

37. J. Adler, *Science*, 166, 1588–1597 (1969).

p. 197

38. C. M. Child, *Patterns and Problems of Development* (University of Chicago Press, Chicago, Ill., 1941).

39. The earlier work is reviewed by L. G. Barth, *Biological Reviews*, 15, 405–420 (1940).

40. *Ibid.*

41. J. A. Miller, *Biological Bulletin*, 73, 369 (1937).

p. 198

42. *Ibid.*, 99, 361–363 (1950).

p. 199

43. S. M. Rose and F. Rose, *Physiological Zoology*, 14, 328–343 (1941); see also S. M. Rose, *Regeneration* (Appleton-Century-Crofts, New York, 1970).

44. W. W. Schwabe, *Symposia of the Society for Experimental Biology*, 25, 301–322 (1971).

45. I. D. J. Phillips, in M. B. Wilkins, ed., *The Physiology of Plant Growth and Development* (McGraw-Hill, New York, 1969); P. Champagnat, *Encyclopedia of Plant Physiology* (Springer-Verlag,

New York, 1965), vol. 15/1, pp. 1106–1164.

46. Reviewed in A. Kühn, *Lectures in Developmental Physiology*,   p. 201
(Springer-Verlag, New York, 1971), pp. 166–200.

47. W. P. Jacobs, in D. J. Carr, ed., *Plant Growth Substances, 1970*   p. 202
(Springer-Verlag, Berlin, 1972), pp. 701–709.

48. Y. P. Chang and W. P. Jacobs, *Plant Physiology, 50,* 635–639
(1972).

49. See A. B. Pardee, *Science, 162,* 632–637 (1968).

50. W. R. Loewenstein, in M. Locke, ed., *The Emergence of Order*   p. 203
in Developing Systems (Academic Press, New York, 1968), pp. 151–183.

51. P. A. Lawrence, *Journal of Experimental Biology, 44,* 607–620
(1966).

52. This suggestion came from a lecture by O. K. Wilby (April
1972).

53. Reviewed by V. C. Twitty, *Growth Symposium, 9,* 133–161   p. 207
(1949).

54. H. C. Dalton, *Journal of Experimental Zoology, 115,* 151–174
(1950).

55. D. A. Ede, *Symposia of the Society for Experimental Biology,*
25, 235–254 (1971).

56. J. T. Bonner, *Morphogenesis* (Princeton University Press,   p. 208
Princeton, N. J., 1952).

57. D. Cohen, *Nature, 216,* 246–248 (1967).   **p. 209**

58. G. E. Fogg, *Annals of Botany,* n.s., *13,* 241–259 (1949).   p. 212

59. For illustrations of the variety of division patterns see E. Fryn-
Claessens and W. Van Cotthem, *Botanical Review, 39,* 71–138 (1973).

60. S. T. Degnen and A. Newton, *Journal of Molecular Biology, 64,*
671–680 (1972).

61. A. Newton, *Proceedings of the National Academy of Sciences,*
U.S., 69, 447–451 (1972).

62. For example, J. B. Angevine and R. L. Sidman, *Nature, 192,* 766–   p. 214
768 (1961). For a more recent study showing the role of cell adhesion
in tissue cultures of brain cells see G. R. DeLong, *Developmental Bi-*
ology, 22, 563–583 (1970).

63. For a review see T. C. Schneirla, *Army Ants* (W. H. Freeman,   p. 216
San Francisco, 1971).

64. E. O. Wilson, *The Insect Societies* (Harvard University Press,
Cambridge, Mass., 1971), pp. 249–258; also Wilson, in E. Sond-
heimer and J. B. Simeone, eds., *Chemical Ecology* (Academic Press,
New York, 1970), pp. 133–155.

9. The Control of Pattern

p. 219   1. This point is illustrated in a delightful way in the front matter of F. W. Went and K. V. Thimann's classic, but now outdated book, *Phytohormones* (Macmillan, New York, 1937); they quote the old nursery rhyme:

"Oats, peas, beans and barley grow,
Oats, peas, beans and barley grow,
Can you or I, or anyone know
How oats, peas, beans and barley grow"

p. 221   2. L. F. Jaffee, *Advances in Morphogenesis, 7,* 295–328 (1968).
3. L. F. Jaffee, *Science, 123,* 1081–1082 (1956).

p. 223   4. For example, L. C. Beadle and F. A. Booth, *Journal of Experimental Biology, 15,* 303–326 (1938).

5. M. S. Steinberg, *Journal of Experimental Zoology, 127,* 1–26 (1954).

p. 224   6. A. M. Turing, *Transactions of the Royal Society,* London, ser. B, *237,* 37–72 (1952).

7. E. F. Keller and L. A. Segel, *Nature, 227,* 1365–1366 (1970); *Journal of Theoretical Biology, 26,* 399–415 (1970).

p. 225   8. Bonner *et al., Developmental Biology, 20,* 72–87 (1969); E. W. Samuel, *ibid., 3,* 317–335 (1961).

9. G. Gerisch, D. Malchow, V. Riedel, E. Muller, and M. Every, *Nature, 235,* 90–92 (1972).

10. J. T. Bonner and M. R. Dodd, *Biological Bulletin, 122,* 13–24 (1962).

11. A. J. Kahn, *Developmental Biology, 18,* 149–162 (1968); reviewed in J. LoBue and A. S. Gordon, eds., *Hormonal Control of Growth and Differentiation* (Academic Press, New York, 1973), chap. 19.

p. 227   12. Reviewed in C. M. Child, *Patterns and Problems of Development* (University of Chicago Press, Chicago, Ill., 1941).

p. 228   13. S. F. Light, *Quarterly Review of Biology, 17,* 312–326 (1942); *18,* 46–63 (1943).

p. 230   14. M. Lindauer, *Communication among Social Bees* (Harvard University Press, Cambridge, Mass., 1961); see also E. O. Wilson, *The Social Insects* (Harvard University Press, Cambridge, Mass., 1971), pp. 174–177.

15. Wilson, *The Social Insects,* pp. 342–343.

16. C. F. Robinow, *Journal of Cellular Biology*, *17*, 123–152   p. 231
(1963).

17. D. R. Stadler, *Journal of Cellular and Comparative Physiology*,   p. 232
39, 449–474 (1952); *Biological Bulletin*, *104*, 100–108 (1953).

18. Reviewed by L. Machlis and E. Rawitscher-Kunkel in C. B.   p. 233
Metz and A. Monroy, eds., *Fertilization* (Academic Press, New York,
1967), vol. 1, pp. 117–161; see also L. Machlis in G. C. Ainsworth
and A. S. Sussman, eds., *The Fungi* (Academic Press, New York,
1966), vol. 2, pp. 415–433.

19. H. A. de Bary, *Comparative Morphology and Biology of the*   p. 236
Fungi, Mycetozoa and Bacteria (Clarendon Press, Oxford, 1887).

20. J. T. Bonner, K. K. Kane, and R. Levey, *Mycologia*, *48*, 13–19   p. 237
(1955).

21. H. Borriss, *Planta*, *22*, 28–69 (1934).

22. L. E. Hawker, *Physiology of Fungi* (University of London
Press, London, 1950).

23. T. Urayama, *The Botanical Magazine*, Tokyo, *69*, 298–299
(1956).

24. H. Hagimoto and M. Konishi, *The Botanical Magazine*, *72*, 359–
366 (1959).

25. See L. von Bertalanffy and J. H. Woodger, *Modern Theories of*
Development (Oxford University Press, London, 1933).

26. K. B. Raper, *Journal of the Elisha Mitchell Scientific Society*,   p. 239
56, 241–282 (1940).

27. B. M. Shaffer, *Quarterly Review of Microscopical Science*, *98*,   p. 240
377–392 (1957).

28. J. T. Bonner, *Journal of Experimental Zoology*, *110*, 259–271
(1949).

29. For a review see J. T. Bonner, *The Cellular Slime Molds* (2nd
ed.; Princeton University Press, Princeton, N. J., 1967).

30. *Ibid.*   p. 241

31. J. T. Bonner and E. B. Frascella, *Journal of Experimental*
Zoology, *121*, 561–571 (1952), esp. p. 569.

32. J. H. Gregg, *Developmental Biology*, *12*, 377–393 (1965).

33. L. Wolpert, *Current Topics in Developmental Biology*, *6*, 183–
224 (1971).

34. J. T. Bonner and M. K. Slifkin, *American Journal of Botany*, *36*,
727–734 (1949).

35. J. T. Bonner, *Proceedings of the National Academy of Sciences*,
U.S., *65*, 110–113 (1970).

p. 242   36. J. T. Bonner, *Quarterly Review of Biology*, 32, 232–246 (1957). For an interesting new hypothesis on the problem of the control of this division line, see D. McMahon, *Proceedings of the National Academy*, U.S., 70, 2396–2400 (1973).

p. 246   37. M. H. Braverman, *Journal of Embryology and Experimental Morphology*, 11, 239–253 (1963); M. H. Braverman and R. G. Schrandt, *Symposia of the Zoological Society of London*, 16, 169–198 (1966).

38. F. Peebles, Roux's *Archiv für Entwicklungsmechanik*, 10, 435–488 (1900).

39. Braverman and Schrandt, *Symposia;* L. J. Hale, in preparation.

p. 248   40. R. D. Campbell, *Developmental Biology*, 15, 487–502 (1967); *Journal of Morphology*, 121, 19–28 (1967); *Journal of Experimental Zoology*, 164, 379–386 (1967); S. Shostak and D. R. Kankel, *Developmental Biology*, 15, 451–463 (1967); S. G. Clarkson and L. Wolpert, *Nature*, 214, 780–783 (1967).

41. L. J. Hale, *Journal of Embryology and Experimental Morphology*, 12, 517–538 (1964).

42. For an up-to-date review see M. Hamburgh, *Theories of Differentiation* (Edward Arnold, London, 1971).

p. 250   43. Reviewed in R. K. Gaze, *The Formation of Nerve Connections* (Academic Press, New York, 1970), and M. Jacobson, *Developmental Neurobiology* (Holt, Rinehart, and Winston, New York, 1970).

44. Jacobson, *Developmental Neurobiology.*

45. Reviewed in R. G. Harrison, *Transactions of the Connecticut Academy of Sciences*, 36, 277–330 (1945).

p. 251   46. Gaze, *The Formation of Nerve Connections;* R. M. Gaze and M. J. Keating, *Nature*, 237, 375–378 (1972).

47. J. Johnson, manuscript in preparation. See also M. Cohen in O. J. Plescia and W. Braun, eds., *Nucleic Acids in Immunology* (Springer-Verlag, New York, 1968), pp. 671–715.

p. 252   48. Reviewed in S. Puiseux-Dao, *Acetabularia and Cell Biology* (Springer-Verlag, New York, 1970).

p. 254   49. V. Tartar, *Growth*, Symposium 5, 21–40 (1941).

50. N. de Terra, *Symposia of the Society for Experimental Biology*, 24, 345–368 (1970).

p. 255   51. Reviewed in V. Tartar, *The Biology of Stentor* (Pergamon Press, London, 1961).

52. A. S. G. Curtis, *Journal of Embryology and Experimental Morphology*, 10, 410–422 (1962).

274

Index

Index

Index

Index